MULTI-CRITERIA DECISION MODELLING

Science, Technology, and Management Series

Series Editor: J. Paulo Davim, Professor

Department of Mechanical Engineering, University of Aveiro, Portugal

This book series focuses on special volumes from conferences, workshops, and symposiums, as well as volumes on current topics of interest in all aspects of science, technology, and management. The series will discuss topics such as mathematics, chemistry, physics, materials science, nanosciences, sustainability science, computational sciences, mechanical engineering, industrial engineering, manufacturing engineering, mechatronics engineering, electrical engineering, systems engineering, biomedical engineering, management sciences, economical science, human resource management, social sciences, engineering education, etc. The books will present principles, model techniques, methodologies, and applications of science, technology, and management.

Optimization Using Evolutionary Algorithms and Metaheuristics
Edited by Kaushik Kumar and J. Paulo Davim

Integration of Process Planning and Scheduling
Approaches and Algorithms
Edited by Rakesh Kumar Phanden, Ajai Jain, and J. Paulo Davim

Understanding CATIA
A Tutorial Approach
Edited by Kaushik Kumar, Chikesh Ranjan, and J. Paulo Davim

Manufacturing and Industrial Engineering
Theoretical and Advanced Technologies
Edited by Pakaj Agarwal, Lokesh Bajpai, Chandra Pal Singh, Kapil Gupta, and J. Paulo Davim

Multi-Criteria Decision Modelling
Applicational Techniques and Case Studies
Edited by Rahul Sindhwani, Punj Lata Singh, Bhawna Kumar, Varinder Kumar Mittal, and J. Paulo Davim

For more information about this series, please visit: https://www.routledge.com/Science-Technology-and-Management/book-series/CRCSCITECMAN

MULTI-CRITERIA DECISION MODELLING

Applicational Techniques and Case Studies

Edited by
Rahul Sindhwani
Punj Lata Singh
Bhawna Kumar
Varinder Kumar Mittal
J. Paulo Davim

CRC Press
Taylor & Francis Group
Boca Raton London New York

CRC Press is an imprint of the
Taylor & Francis Group, an **informa** business

First edition published 2022
by CRC Press
6000 Broken Sound Parkway NW, Suite 300, Boca Raton, FL 33487-2742

and by CRC Press
2 Park Square, Milton Park, Abingdon, Oxon, OX14 4RN

CRC Press is an imprint of Taylor & Francis Group, LLC

Library of Congress Cataloging-in-Publication Data
Names: Sindhwani, Rahul, editor.
Title: Multi-criteria decision modelling : applicational techniques and case studies / edited by Rahul Sindhwani, Punj Lata Singh, Bhawna Kumar, Varinder Kumar Mittal, and J. Paulo Davim.
Other titles: Multicriteria decision modelling
Description: First edition. | Boca Raton, FL : CRC Press, 2021. |
Series: Science, technology, and management |
Includes bibliographical references and index.
Identifiers: LCCN 2021006148 (print) | LCCN 2021006149 (ebook) |
ISBN 9780367645588 (hbk) | ISBN 9781003125150 (pbk) | ISBN 9780367645649 (ebk)
Subjects: LCSH: Multiple criteria decision making. |
Decision support systems. | Engineering design--Data processing--Case studies. |
Management science--Data processing--Case studies.
Classification: LCC T57.95 .M8184 2021 (print) | LCC T57.95 (ebook) |
DDC 658.4/03--dc23
LC record available at https://lccn.loc.gov/2021006148
LC ebook record available at https://lccn.loc.gov/2021006149

ISBN: 978-0-367-64558-8 (hbk)
ISBN: 978-0-367-64564-9 (pbk)
ISBN: 978-1-003-12515-0 (ebk)

Typeset in Times
by MPS Limited, Dehradun

Contents

Preface....................vii
Editors....................ix
Contributors....................xiii

Chapter 1 Structural Modelling for the Factors of Facility Management in the Healthcare Industry Using the TISM Approach....................1

Rahul Sindhwani, Punj Lata Singh, Rajender Kumar, Bhawna Kumar, Varinder Kumar Mittal, Sumeet Dixit, and Anil Jindal

Chapter 2 Assessment of Optimum Automotive Brake Friction Formula by Entropy-VIKOR Technique....................23

Vishal Ahlawat, Parinam Anuradha, Sunil Nain, Sanjay Kajal, Praveen Tewatia, and Mukesh Kumar

Chapter 3 Ranking of Phase Change Materials for Medium Temperature Thermal Energy Accumulation System Using Shannon Entropy, TOPSIS, and VIKOR Methods....................43

Ankit Yadav, Vikas, and Sushant Samir

Chapter 4 Identification and Analysis of Enablers of Socially Sustainable SMEs Using the MCDM Approach....................65

Shubhanshi Mittal, Shreya Gupta, and Vernika Agarwal

Chapter 5 Application of the MCDM Technique for an Astute Decision-Making Process in the Renewable Energy Sector....................83

Eshan Bajal, Alakananda Chakraborty, Muskan Jindal, and Shilpi Sharma

Chapter 6 Assessing Challenges in the Implementation of Sustainable Human Resource Management Using the Fuzzy DEMATEL Approach....................107

Vernika Agarwal, Snigdha Malhotra, and Tilottama Singh

Chapter 7 A Novel Entropy Measure for Linguistic Intuitionistic Fuzzy Sets and Their Application in Decision-Making 121

Kamal Kumar, Naveen Mani, Amit Sharma, and Reeta Bhardwaj

Chapter 8 An Insight into Decision-Making Using the Fuzzy Analytic Hierarchy Process (FAHP) 139

Nikhil Dev and Rajeev Saha

Chapter 9 Application of the Fuzzy Analytic Hierarchy Process for Solar Panel Vendor Selection 155

Rajeev Saha and Nikhil Dev

Chapter 10 Using Multi-Criteria Decision-Analysis to Build a Long-Term Business Strategy 179

Sandeep Bhasin

Chapter 11 Systematic Review and Deliberation of Various Multi-Criteria Decision-Making Techniques 189

Muskan Jindal and Areeba Kazim

Index 205

Preface

'Making good decisions is a crucial skill at every level' said Peter F. Drucker (1909–2005), the Marketing Guru. In the present scenario, the single-point agenda is used to handle and manage real-life problems within the environment by thoroughly examining and analyzing the decisions before their implementation. Decision-making is quite complex especially when it is based on multiple criteria. The general perception regarding decision-making is that every decision has a limitation that constraints implementation. The presence of several intervening variables and their dependencies (inter-relationships) might make it difficult to find potential solutions as there are diverse objectives envisioned for a project. The availability of enormous amounts of information can render rationalizing decision-making extremely difficult which adds to its complexity.

It is a well-acknowledged fact that the process of decision-making is subjective in nature and thus, it is important for decision-makers to choose the best fitting solutions while making rational decisions. Consequently, the role of decision-makers during any volatile environment becomes more crucial, as there is not enough historical data to understand the pattern of key indicators in real-life scenarios. Henceforth, the study of optimization of multi-criteria decision-making (MCDM) tools in this book will provide the necessary solutions to researchers, policymakers, managers, and graduate students. This book reveals the attempts made by researchers to discover the diverse tools which are useful for strategic and effective decision-making in real-life problems, especially the ones involving multiple criteria. The main emphasis of this book is on laying the practical foundations using MCDM models which act as guides in addressing the emerging industrial/social issues and obtaining the optimal product/process utilities that lead to sustainability.

Editors

Rahul Sindhwani earned a PhD in Industrial Engineering and Management in 2017, an MTech in Mechanical Engineering with distinction and a gold medal in 2010, a BTech in Mechanical Engineering in 2008 at the Kurukshetra University, India. He is an Assistant Professor in the Mechanical Engineering Department, Amity University, Uttar Pradesh, India. He is an active lifetime member of the Indian Society of Technical Education and a fellow member of the Institution of Engineering and Technology. He has more than 10 years' worth of experience in teaching, research, and administration experience. He has contributed more than 25 research papers and book chapters in the area of industrial engineering and management in various reputable national and international journals. His area of expertise is industrial engineering and management, supply chain management, lean, green, and agile manufacturing systems, along with analysis of problems using various optimization techniques.

Punj Lata Singh is an Assistant Professor in the Department of Civil Engineering (CE), Amity School of Engineering and Technology (ASET), Amity University Uttar Pradesh (AUUP), Noida, India. She earned an MTech in Fluid Engineering at the MNNIT, Allahabad, India in 2010 and a BTech in Mechanical Engineering at the UPTU, Lucknow, India in 2006. She has more than 10 years' worth of experience in teaching, research, and administration experience. She is an active lifetime member of the Indian Society of Technical Education. As an academician, her primary research interests are engineering and technology. She has been actively associated with organizing and attending several national and international conferences and published many research papers on the subjects of thermal engineering, industrial engineering, and management in various international journals that are renowned in the fields of fluids engineering, industrial engineering, and management. She has worked in the field of implementation, analysis, and optimization of problems using multi-criteria decision modelling techniques and various experimentation in the thermal area.

Bhawna Kumar is a distinguished academician, researcher, trainer, and management consultant with a brilliant career in education. The gold medalist of her batch of BE at the Delhi University, she was awarded the gold medal in her MBA programme. She has specialized in instrumentation and control engineering and management and has more than 26 years' worth of work experience covering the areas of strategic management, project execution, project management, ICT, consultancy, research, teaching, and training. She is a Senior Professor of Strategy and a Vice President of the Amity Education Group who is closely associated with the strategic initiatives in the areas of training, consulting, and internationalization of all of its institutions. An ardent researcher, she has been actively involved in organizing several national and international conferences and has published numerous research papers in the areas of IT and management. Professor Kumar has been instrumental in furthering the vision of the internationalization of Amity and has contributed to the launch of the many global campuses of Amity, including London, Singapore, Mauritius, Uzbekistan, New York, and San Francisco, while also nurturing the International Business School, Amity Greater Noida Campus, Study Abroad, and three continent and international programs. Moreover, she has extensive experience in addressing various forums, institutions, and corporates, such as Maruti, Jones Lang LaSalle, PGCIL, Alchemist, etc. Professor Kumar is also a member of many prestigious editorial boards.

Varinder Kumar Mittal is a manufacturing engineer at Dutch Industries Ltd. Pilot Butte SK Canada. For 16 years, he has served on the faculties of many departments of mechanical engineering at various engineering institutes in India. He is a graduate in Mechanical Engineering from the Punjab Technical University, Jalandhar, India in 1999, a postgraduate in Mechanical Engineering with a Specialization in Production Engineering from the Punjab Technical University, Jalandhar, India in 2004, and a PhD in Green Manufacturing from BITS Pilani, India in 2014. His teaching and research interests are in the field of manufacturing engineering and management along with analysis of problems using structural equation modelling, statistical analysis, interpretive structural modelling, multi-criteria decision modelling, etc. He has worked in the manufacturing industry for more than five years.

J. Paulo Davim earned a PhD in Mechanical Engineering in 1997, an MSc in Mechanical Engineering (Materials and Manufacturing Processes) in 1991, a five-year Mechanical Engineering Degree at the University of Porto (FEUP) in 1986, the Aggregate Title (Full Habilitation) at the University of Coimbra in 2005, and the DSc at the London Metropolitan University in 2013. He is a Senior Chartered Engineer by the Portuguese Institution of Engineers, with an MBA as well as a specialist title in Engineering and Industrial Management. He is also a Eur Ing by FEANI-Brussels and Fellow (FIET) by IET-London. Dr. Davim is a Professor in the Department of Mechanical Engineering of the University of Aveiro, Portugal. He has more than 30 years of teaching experience and research experience in manufacturing, materials, and mechanical and industrial engineering, with a special emphasis in machining and tribology. His interests include management, engineering education, and higher education for sustainability. He has guided several postdoctoral, PhD, and master's students, as well as coordinated and participated in several financed research projects. He has received considerable scientific awards. Dr. Davim has worked as an evaluator of projects for ERC-European Research Council and other international research agencies, as well as an examiner of PhD theses for countless universities in different countries. He is the editor-in-chief of several international journals, guest editor of journals, books editor, book series editor, and the scientific advisor for many international journals and conferences. Presently, he is an editorial board member of 30 international journals and acts as the reviewer for more than 100 prestigious Web of Science journals. In addition, he has also published, as editor and as co-editor, more than 125 books and, as author and co-author, more than 10 books, 80 book chapters, and 400 articles in journals and conferences (more than 250 articles in journals indexed in Web of Science core collection/h-index 54+/9000+ citations, SCOPUS/h-index 58+/11000+ citations, Google Scholar/h-index 75+/18500+).

Contributors

Vernika Agarwal
Amity International Business School
Amity University
Noida (UP), India

Vishal Ahlawat
Department of Mechanical Engineering
University Institute of Engineering and Technology
Kurukshetra (Hr.), India

Parinam Anuradha
Department of Mechanical Engineering
University Institute of Engineering and Technology
Kurukshetra (Hr.), India

Eshan Bajal
Department of Computer Science and Engineering
Amity University
Noida (UP), India

Reeta Bhardwaj
Department of Mathematics
Amity University
Gurugram (Hr.), India

Sandeep Bhasin
Amity International Business School
Amity University
Noida (UP), India

Alakananda Chakraborty
Department of Computer Science and Engineering
Amity University
Noida (UP), India

Nikhil Dev
Department of Mechanical Engineering
J.C. Bose University of Science and Technology
Faridabad (Hr.), India

Sumeet Dixit
Department of Mechanical Engineering
Amity University
Noida (UP), India

Shreya Gupta
Department of Business Studies
Amity University
Noida (UP), India

Anil Jindal
Department of Mechanical Engineering, Giani Zail Singh Campus College of Engineering and Technology
Bathinda (Pb.), India

Muskan Jindal
Department of Computer Science and Engineering
Amity University
Noida (UP), India

Sanjay Kajal
Department of Mechanical Engineering
University Institute of Engineering and Technology
Kurukshetra (Hr.), India

Areeba Kazim
Department of Computer Science and Engineering Noida Institute of Engineering and Technology
Greater Noida (UP), India

Kamal Kumar
Department of Mathematics
Amity University
Gurugram (Hr.), India

Mukesh Kumar
Department of Mechanical Engineering State Institute of Engineering and Technology
Nilokheri (Hr.), India

Rajender Kumar
Department of Mechanical Engineering FET, Manav Rachna International Research and Studies
Faridabad (Hr.), India

Snigdha Malhotra
Department of Business Studies
Amity University
Noida (UP), India

Naveen Mani
Department of Mathematics
Chandigarh University
Mohali (PB), India

Shubhanshi Mittal
Amity International Business School
Amity University
Noida (UP), India

Varinder Kumar Mittal
Dutch Industries Ltd.
Pilot Butte, SK, Canada

Sunil Nain
Department of Mechanical Engineering
University Institute of Engineering and Technology
Kurukshetra (Hr.), India

Rajeev Saha
Department of Mechanical Engineering
J.C. Bose University of Science and Technology
Faridabad (Hr.), India

Sushant Samir
Department of Mechanical Engineering Punjab Engineering College
Chandigarh, India

Amit Sharma
Department of Mathematics
Amity University Haryana
Gurugram (Hr.), India

Shilpi Sharma
Department of Computer Science and Engineering
Amity University Noida (UP), India

Rahul Sindhwani
Department of Mechanical Engineering
Amity University
Noida (UP), India

Bhawna Kumar
Amity Education Group
Amity University
India

Punj Lata Singh
Department of Civil Engineering
Amity University
Noida (UP), India

Tilottama Singh
Department of Business Studies
Amity University
Noida (UP), India

Praveen Tewatia
Department of Mechanical Engineering
University Institute of Engineering and Technology
Kurukshetra (Hr.), India

Vikas
Department of Mechanical Engineering
Punjab Engineering College
Chandigarh, India

Ankit Yadav
Department of Mechanical Engineering
Punjab Engineering College
Chandigarh, India

1 Structural Modelling for the Factors of Facility Management in the Healthcare Industry Using the TISM Approach

Rahul Sindhwani, Punj Lata Singh, Rajender Kumar, Bhawna Kumar, Varinder Kumar Mittal, Sumeet Dixit, and Anil Jindal

CONTENTS

1.1 Introduction 2
1.2 Identification of the Factors Affecting Facility Management in the Healthcare Sector 2
1.2.1 Organisational Structure Design 3
1.2.2 Talent Management 3
1.2.3 Conflict Management 3
1.2.4 Inter-Connectivity Management 4
1.2.5 Enterprise-Wide Integration of Learning 4
1.2.6 Customer Feedback Management 4
1.2.7 Market Awareness 4
1.2.8 External Collaborations 5
1.2.9 IT Tools/Technologies Adoption 5
1.2.10 Patient Assumption and Hospital Discernment 5
1.3 Methodology 5
1.3.1 Identified Factors Affecting Facility Management in Healthcare Organisations 6
1.3.2 TISM Analysis 7
1.4 MICMAC Analysis 10
1.4.1 The Development of the MICMAC Graph 10
1.4.2 Development of MICMAC Rank 11
1.5 Result and Discussion 11
1.6 Conclusion 14
References 15

1.1 INTRODUCTION

In industrial terminology, it is important to sustain industry contribution in the market in the long-term scenario (Kumar et al., 2017a; Shanker et al., 2019; Sindhwani et al., 2018). In order to do this, timely modification or updating of existing facilities, processes, and products are requisite (Sindhwani, Singh, Chopra, et al., 2019). In this regard, most industries are opting for agile manufacturing, i.e. introducing facility management in their industrial operations. The term 'facility management' refers to the adaptation of advanced tools and techniques used to satisfy consumer needs (Naylor et al., 1999). Facility management helps in increasing the responsiveness of the industry based on consumer demand prediction and aids in preparing the necessary strategy to respond accordingly (Yusuf et al., 1999).

While going through the literature, it seems difficult to decide on modifications or make amendments to existing facilities because there are a lot of attributes associated with these kinds of decisions. For making such decisions, the decision-maker has to simultaneously consider multiple attributes, which itself is a huge task. To simplify this, the multi-criteria decision-making approach is suggested. In the MCDM approach, multiple attributes (enablers/factors or barriers) are analyzed in order to reveal the best alternative to go with most, based on the various set criteria. It is an extensive methodology and is used in almost all kinds of industries in the present-day scenario.

The current work has aimed to discover the solution for facility management in the healthcare sector through the MCDM approach. With the use of the MCDM tool, the road map is being developed for healthcare organisations, which further helps in strategizing the successful adaptation of prioritized factors (outcomes of the study) for facility management of the healthcare sector. The present study identified 10 factors associated with healthcare sector facility management through an extensive literature review. Using the TISM technique, the interdependence of the factors has been identified, and these are accordingly ranked using the MICMAC analysis.

1.2 IDENTIFICATION OF THE FACTORS AFFECTING FACILITY MANAGEMENT IN THE HEALTHCARE SECTOR

Worldwide, healthcare has become a prime concern for human beings. Everyone is now looking for better healthcare facilities, such as safe and hygienic services (Kumar et al., 2017c; Sindhwani, Singh, Iqbal, et al., 2019). The services offered to patients vary based on attributes such as service provider capability, patient financial conditions, disease criticality, *etc.* Now, even patients from the higher class are expressing concerns with record-keeping for pre- and post-medical history related to the treatment offered to them (Kumar et al., 2019). In order to manage all kinds of facilities in the healthcare industry, it is paramount to offer mixed services. Moreover, for such services, the industry must have the ability to meet all consumer demands without compromising the prime objectives, i.e. patient care with the utmost consideration for safety and hygiene. In this regard, facility management in operations of healthcare organisations helps in overcoming the target challenges, such as less expensive treatment, increased availability, safety in all aspects, and efficient healthcare facilities, *etc* (Sindhwani & Malhotra, 2017). Even if facility management provides a fruitful contribution to the healthcare industry, there are still a few factors associated with this sector that prevents its successful implementation.

The literature reveals that most of the healthcare sector units are working in a decentralized manner. Decentralization in operations encourages innovation in terms of finding and grabbing opportunities. Such kinds of opportunities further help the organisation to scale up the business by offering diversity in products and services. However, there are numerous organisations that do not benefit from decentralization because of reasons such as lack of commitment, diversity in operations, social cohesion, *etc*. Therefore, it became paramount to introduce facility management among all the operational units of the healthcare organisation. Facility management confirms the activities done by various operational units of healthcare organisations in such a manner that everyone involved would feel pleasant and benefitted. The present section of the study aims to discuss the various factors present in facility management in a healthcare organisation, as explained below:

1.2.1 Organisational Structure Design

Despite the kind of operations performed within and outside of the organisation, it is crucial for the organisational structure to support all kinds of existing operations. Furthermore, while designing and developing the organisational structure, it is important to consider some space for future advancement. A stricter organisation has an influence on the facility management of a healthcare group. For a well-designed organisational structure, synchronization among various operations is also a complex task, and this has a direct impact on healthcare organisation efficiency and effectiveness.

1.2.2 Talent Management

In industrial terminology, the employees play an important role in the growth of an organisation. Similarly in the healthcare sector, it is important to manage the talent (employees) in an efficient manner. Motivated employees always help an organisation to serve beyond its capabilities and showcase flexibility which further contributes to maintaining facility management. In the healthcare sector, multiple skills are requisites for an individual to be able to handle different situations. The taskforce in healthcare attends, checks, and delivers the services to the patients in such a manner that the patient can feel and experience these accordingly. In order to do this, the training of an employee also plays an important role in the employee's welfare and capability enhancement. Training is provided to all of the employees to update them on advanced tools and techniques just before these new technologies are imparted to the organisation (Yusuf et al., 1999).

1.2.3 Conflict Management

In organisations, a difference between management objectives and employees' working styles would lead to the failure of an organisation. Facility management of a healthcare organisation relies on the coherence among the management and employees. There should be a strategy coherence between the management and the employees in order to work out all of the guidelines before these are executed. This would lead to benefits for both parties and, moreover, the employees' criticisms on any issue are resolved in a timely manner (Vinodh & Devadasan, 2011).

1.2.4 Inter-Connectivity Management

In any organisation, there should be defined boundaries for all of the operational units of the organisation so that the activities happening under different units are unaffected. Further, these boundaries reveal the interdependency among the activities occurring under the various operational units. This is also necessary in order to measure the performance of an individual unit and its impact on others. (Sherehiy et al. 2007). For organisations where boundaries are not defined, there seems to be a lack in their performance, and this also influences facility management.

1.2.5 Enterprise-Wide Integration of Learning

Information and knowledge are the two key points in the development of all kinds of organisations, especially in the healthcare sector (Gehani, 1995). Information regarding patient details and care history must be known by anyone concerned with patient care (Stelson et al., 2017). This gives pleasure to patients when they are availing of services from concerned parties. (Kidd, 1995). Integration of all of the information is of utmost importance in order for the organisation to help manage and improve the healthcare services.

1.2.6 Customer Feedback Management

Customer feedback is a very valuable piece of information for any organisation. The true opinion of a patient on the treatment given can help the organisation to improve patient experience, and, at the same time, this will also help the organisation to analyze its working procedures (Vinodh et al., 2012). It is essential for a healthcare organisation to always strive towards providing the best experience for the patients as well as their speedy recovery, because this plays a key role in the treatment of the patients.

1.2.7 Market Awareness

The global market is unpredictable and is always in dynamic mode with respect to variations in demand versus supply. Hence, it is pertinent to know the need of current consumers so that they will offer services according to requisitions. For the healthcare sector, market awareness among the consumers (patients) should be of utmost priority. Awareness regarding facilities and services directly reveals organisational capability and specialization (Christopher & Towill, 2001).

1.2.8 External Collaborations

In healthcare organisations, small-scale institutions are always outsourcing services and products offered by other firms and thereby gain a dependency on them. A healthy networking relationship among all of the outsourcing partners is requisite, but this creates a hurdle in the growth of an individual organisation. This factor also affects the organisation's facility management by distracting the value chain of healthcare organisations.

1.2.9 IT Tools/Technologies Adoption

In facility management formation in an organisation, the flow of information is one of the most important tasks that need to be prioritized. In order to do this, information technology-enabled systems are currently introduced and able to contribute to organisational growth. Although, this factor affects healthcare organisations because different hospitals have now designated a particular slot to decide the budget for the purchase or maintain these IT-enabled systems. A proper plan should be prepared before selecting the IT-enabled system so that the industry can attain maximum benefits.

1.2.10 Patient Assumption and Hospital Discernment

In marketing science, it is well-known that one cannot predict consumer needs perfectly. There must be certain assumptions regarding the patient care system, and organisations are working on them. Still, a few patients demand customized services, which are mostly not possible all of the time. Facility management among the operational units is required to meet consumer demands. Assumptions should be reframed after a fixed period, and the response must be prepared to properly clear all doubts related to the patient's health (Talib & Rahman, 2015).

1.3 METHODOLOGY

In day-to-day life, everyone has to make decisions. Likewise, in order to perform operations smoothly in an organisation, it is pertinent to make the right decision at the right time. However, it becomes difficult to do this when the alternatives increase in number. In order to determine the best feasible solution based on a few set criteria, the MCDM tool is used to define the best alternatives. The various MCDM tools in use are the AHP, ANP, TOPSIS, ISM, TISM, *etc.*

At present, the TISM approach is employed, and this is generally used for analyzing data with multiple attributes. It is the modified approach to the existing version of analysis known as ISM, developed by Warfield (1974). ISM was used for finding solutions to complex and intricate problems. Though ISM is a useful technique of analysis, it also has shortcomings. It does not mention the operation of direct links that creates a relation between different factors. Moreover, it does not consider any need for the creation of transitive links. There are many studies where TISM is used as an analysis tool. For this study, it is used as the analysis tool because it helps in the creation of meaningful links among the factors under consideration. The MICMAC analysis, which is used along with the TISM approach, makes it easier for the reader and analyst to clearly outline the factors which should be given priority on the basis of their driving power which, in turn, shows how much they affect the other factors. The various steps involved in the analysis are described in a stepwise method, along with the necessary tabular and graphical information.

1.3.1 Identified Factors Affecting Facility Management in Healthcare Organisations

Factors have been identified based on the study of various literature reviews and expert opinions. These factors are recognized as the most prominent ones through the studies. The factors identified are depicted in Table 1.1.

TABLE 1.1
Factors Affecting Facility Management in an Organisation as Identified in the Literature

	Factors	Description	References
E1	Organisational structure of healthcare industry	Horizontal and flat organisational structure with fewer power differentials, decentralized knowledge, and adherence to authority.	Aravind Raj et al. (2013)
E2	Talent management	Workforce of the organisation should have various skills and be able to tackle multidisciplinary problems.	Christopher (2000)
E3	Conflict management	Cooperation between management and employee.	Vinodh and Devadasan (2011)
E4	Inter-connectivity management	Cooperation between multiple operating units.	Sherehiy et al. (2007)
E5	Enterprise-wide integration of learning	Flexibility in technical and operational setups that can be used across different units.	Yusuf et al. (1999)
E6	Customer feedback management	There should be a proper mechanism to grab the patient's needs.	Vinodh et al. (2012)
E7	Market awareness	Prior to implementing agile factors, one should have proper reach and response from the market.	Guimarães and de Carvalho (2011)
E8	External collaborations	Medicine contracts must be outsourced.	Christopher and Towill (2001)
E9	IT tools/technologies adoption	Healthcare organisations must be well equipped with the latest IT and multimedia technology.	Vinodh et al. (2012)
E10	Patient assumption and hospital discernment	Patients must be well assed and given required medicines.	Talib and Rahman (2015)

1.3.2 TISM Analysis

The TISM analysis of various governing factors includes the following steps:

Step–1 Establishment of the Contextual Relationship: Every factor is checked pair-wise to identify its relationship with its correspondent, for instance, checking whether factor A is related to factor B through some subtle relation or vice versa. These relations have been depicted in Annexure 1.

Step–2 Interpretation of Relationship: If there is a relation between a pair of factors A and B, then it is denoted as 'Yes' (Y) else 'No' (N). This has been show-cased in Table 1.2. The reasoning for 'Y' and 'N' has been provided in Annexure 1.

Step–3 Checking Transitivity and Forming the Initial and Final Reachability Matrix: All the Y and N in the inter-relationship matrix have been changed to 0 and 1 respectively in order to form the initial reachability matrix, as given by Table 1.3. For checking the transitive link, that is, if factor A is related to factor B and in turn, factor B has a direct relation with factor C, then transitivity says that A is related to C. The transitive links are denoted as 1* and are set accordingly in the final reachability matrix. The final reachability matrix is depicted in Table 1.4.

Step–4 Level Partitioning: Level partitioning is done by accumulative grouping the individual elements into three different sets, namely the antecedent set, the reachability set, and the intersection set. The factors that influence other factors form the reachability set. The factors that are being influenced constitute the antecedent set. The overlap of the reachability set with the antecedent set is called the intersection set. Upon the identification of these three sets, the first iteration is done by determining the entries with the same intersection set and reachability set and is designated under the level I. For the second iteration, the factors designated under level I are removed, and the process is carried out. At this point, the factors with the same intersection set and reachability set are designated under level II. This

TABLE 1.2
Inter-Relationship Matrix (*Assertion to the assertive responses is given in Annexure 1.)

Factors	E1	E2	E3	E4	E5	E6	E7	E8	E9	E10
E1	–	Y	Y	Y	Y	Y	N	N	N	N
E2	N	–	Y	Y	Y	Y	Y	Y	Y	Y
E3	N	1	–	Y	Y	Y	N	N	N	N
E4	N	N	N	–	N	Y	N	N	N	N
E5	N	N	N	N	–	N	N	N	1	N
E6	N	N	N	N	N	–	N	N	N	N
E7	N	N	N	N	N	Y	–	N	N	1
E8	N	N	N	N	N	N	N	–	N	N
E9	N	N	N	Y	N	Y	N	N	–	N
E10	N	N	N	N	N	N	N	N	N	–

TABLE 1.3
Initial Reachability Matrix

Factors	E1	E2	E3	E4	E5	E6	E7	E8	E9	E10
E1	–	1	1	1	1	1	0	0	0	0
E2	0	–	1	1	1	1	1	1	1	1
E3	0	1	–	1	1	1	0	0	0	0
E4	0	0	0	–	0	1	0	0	0	0
E5	0	0	0	0	–	0	0	0	1	0
E6	0	0	0	0	0	–	0	0	0	0
E7	0	0	0	0	0	1	–	0	0	1
E8	0	0	0	0	0	0	0	–	0	0
E9	0	0	0	1	0	1	0	0	–	0
E10	0	0	0	0	0	0	0	0	0	–

TABLE 1.4
Final Reachability Matrix

Factors	E1	E2	E3	E4	E5	E6	E7	E8	E9	E10	Driving Power
E1	–	1	1	1	1	1	*1**	*1**	*1**	*1**	10
E2	0	–	1	1	1	1	1	1	1	1	9
E3	0	1	–	1	1	1	*1**	*1**	*1**	*1**	9
E4	0	0	0	–	0	1	0	0	0	0	2
E5	0	0	0	0	–	*1**	0	0	1	0	3
E6	0	0	0	0	0	–	0	0	0	0	1
E7	0	0	0	0	0	1	–	0	0	1	2
E8	0	0	0	0	0	0	0	–	0	0	1
E9	0	0	0	1	0	1	0	0	–	0	5
E10	0	0	0	0	0	0	0	0	0	–	1
Dependence	1	3	3	5	4	8	4	5	5	6	

process is repeated until the last level is reached. Level partitioning and successive iterations are shown in respective Tables 1.4, 1.5, 1.6, 1.7, 1.8, 1.9, and 1.10.

Step–5 Development of the Digraph: Based on the iterations carried out above, a digraph is formed depicting various levels. The digraph is shown in Figure 1.1, along with a detailed description in Figure 1.2. The factors that lie in level I occupy the topmost position, followed by respective levels, with the last being level V as per the number of iterations carried. The digraph also indicates the direct and indirect links corresponding with different levels, as per the final reachability matrix. The direct links correspond to the '1' and indirect links are '1*'.

TABLE 1.5
First Iteration

Factors	Reachability Set	Antecedent Set	Intersection Set	Level
E1	1,2,3,4,5,6,7,8,9,10	1	1	
E2	2,3,4,5,6,7,8,9,10	1,2,3	2,3	
E3	2,3,4,5,6,7,8,9,10	1,2,3	2,3	
E4	4,6	1,2,3,4,9	4	
E5	5,6,9	1,2,3,5	5	
E6	6	1,2,3,4,5,6,7,9	6	I
E7	6,7,10	1,2,3,7	7	
E8	8	1,2,3,8,9	8	I
E9	4,6,8,9,10	1,2,3,5,9	9	
E10	10	1,2,3,7,9,10	10	I

TABLE 1.6
Second Iteration

Factors	Reachability Set	Antecedent Set	Intersection Set	Level
E1	1,2,3,4,5,7,9	1	1	
E2	2,3,4,5,7,9	1,2,3	2,3	
E3	2,3,4,5,7,9	1,2,3	2,3	
E4	4	1,2,3,4,9	4	II
E5	5,9	1,2,3,5	5	
E7	7	1,2,3,7	7	II
E9	4,9	1,2,3,5,9	9	

TABLE 1.7
Third Iteration

Factors	Reachability Set	Antecedent Set	Intersection Set	Level
E1	1,2,3,5,9	1	1	
E2	2,3,5,9	1,2,3	2,3	
E3	2,3,5,9	1,2,3	2,3	
E5	5,9	1,2,3,5	5	
E9	9	1,2,3,5,9	9	III

TABLE 1.8
Fourth Iteration

Factors	Reachability Set	Antecedent Set	Intersection Set	Level
E1	1,2,3,5	1	1	
E2	2,3,5	1,2,3	2,3	
E3	2,3,5	1,2,3	2,3	
E5	5	1,2,3,5	5	IV

TABLE 1.9
Fifth Iteration

Factors	Reachability Set	Antecedent Set	Intersection Set	Level
E1	1,2,3	1	1	
E2	2,3	1,2,3	2,3	V
E3	2,3	1,2,3	2,3	V

TABLE 1.10
Sixth Iteration

Factors	Reachability Set	Antecedent Set	Intersection Set	Level
E1	1,2,3	1	1	VI

1.4 MICMAC ANALYSIS

This analysis is divided into two phases. The first is the preparation of the MICMAC chart. The second is obtaining a MICMAC rank in which driving power and dependence of factors are used (Mittal et al., 2018; 2019). For obtaining the driving power, the elements corresponding with a particular row in the final reachability matrix are added, while dependence is achieved by adding the factors of a particular column interrelated to a respective factor.

1.4.1 The Development of the MICMAC Graph

The classification of factors is done with four different measures:

Autonomous measures: These are the factors whose driving power and dependence are low.

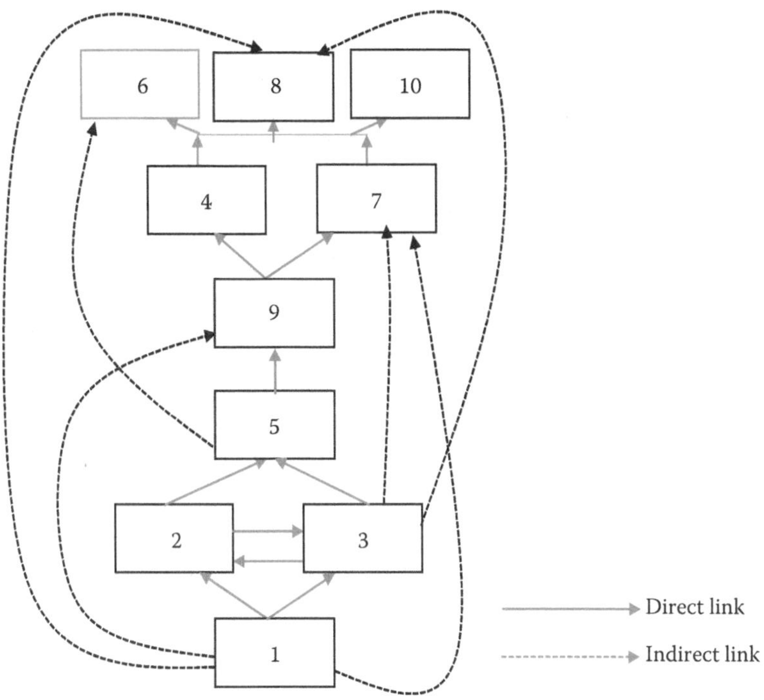

FIGURE 1.1 Diagraph of the Present Study.

Dependent measures: These are the factors whose driving power is low, but dependence is high.
Linkage measures: These are the factors whose driving power and dependence are high.
Independent measures: These are the factors whose driving power is high, but dependence is low.
The development of the MICMAC graph is shown in Figure 1.3.

1.4.2 Development of MICMAC Rank

Rank is assigned to a factor based on its driving power and dependence. The first rank is assigned to the factor with the strongest driving power and least dependence, while the last rank is reserved for the one with the highest dependence. Rank one factors are the most crucial ones, and the rank six factors correspond to the least effective ones. The ranks of respective elements are shown in Table 1.11.

1.5 RESULT AND DISCUSSION

As per the analysis carried in this paper, various factors affecting facility management in the healthcare sector are identified and are tabulated (Table 1.1), and these factors are approached with intensive study of research papers and experts' opinions. Once the

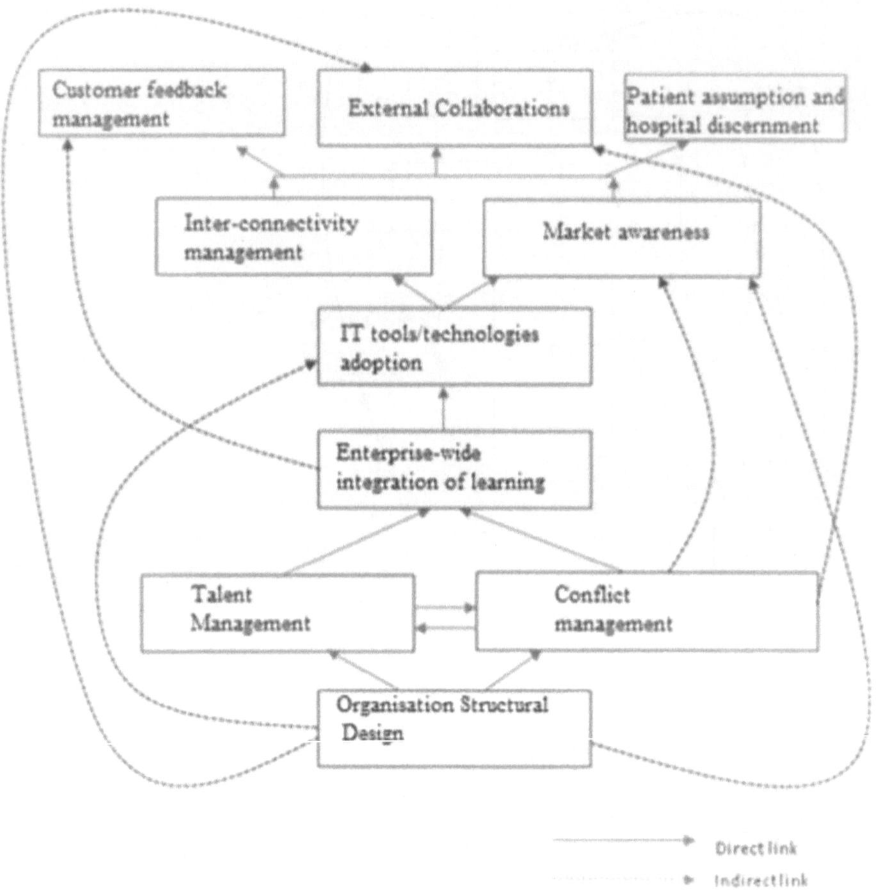

FIGURE 1.2 Diagraph with Detailed Description for Individual Factors.

identification of factors is completed, these are checked for inter- and intra-relationships through the TISM technique, and a final reachability matrix (Table 1.4) is prepared as further several iterations are carried out (as shown in Tables 1.5–1.10) resulting in the formation of a digraph (Figure 1.1) depicting different relevant factors arranged in the particular fashion of their importance; the lower is the foundational or most important factor, and the topmost is the least relevant has the most dependent factors on the former ones. The MICMAC analysis is conducted to reach the final conclusion. The tabulated result dignifies various factors into certain categories of Autonomous factors: E4, E5, E7, E8, and E9; Dependence factors: E6, E10; Linkage factors: null; Driving factors: E1, E2, and E3; thus, the results obtained show that *organisational structure* (E1) is of the utmost importance, and healthcare organisations should work synchronously with all of its units both efficiently and effectively. The next important factor is *a motivated, multi-skilled, and flexible workforce* followed by *the coherence of management and employees* (E2). Workers of the organisation should have multiple skills so that they can handle different situations by making suitable and timely decisions. They should be flexible

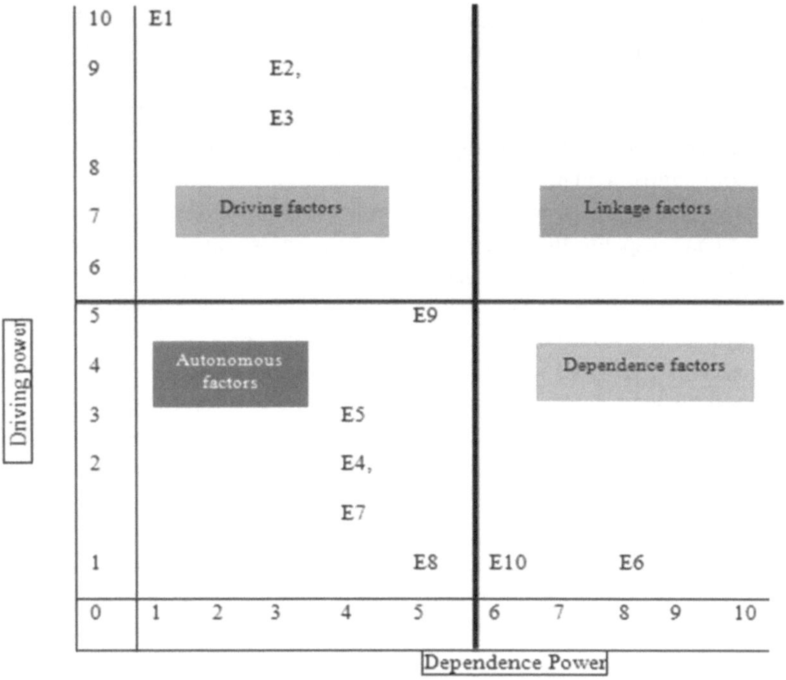

FIGURE 1.3 MICMAC Graph for the Present Work.

TABLE 1.11
Ranking of Factors of Facility Management in the Healthcare Organisation

Factors	Description	Rank
E1	Organisational structure of healthcare industry	1
E2	Talent management	2
E3	Conflict management	2
E5	Enterprise-wide integration of learning	3
E9	IT tools/technologies adoption	4
E4	Inter-connectivity management	5
E7	Market awareness	5
E8	External collaborations	6
E10	Patient assumption and hospital discernment	6
E6	Customer feedback management	6

enough to be able to perform multiple tasks. Furthermore, management and employees should work out certain guidelines which are profitable to both parties. Moreover, employees' issues should be timely resolved by the management, and along with this, *coherence of management and employees* (E3) is of equal importance. *The availability of adequate training for the organisation's workers* (E5) is the next important factor. An update of the technology necessary to handle different processes, and hence, *the adoption of IT technologies* (E9) plays an important role. There should be *loose boundaries among different operation units* (E4), which refers to inter-departmental dependence. This helps in the accomplishment of the assigned work time without the restriction of multiple written formalities. *Market awareness* (E7) is also a critical factor because, prior to implementing agile techniques in the healthcare sector, one should have proper reach and response from the market. The market is unpredictable and is always changing in terms of the variation in demand and supply. The last three important factors in descending order of importance are: *outsourcing* (E8), i.e. medicines and other essentials required for the healthcare sector should be outsourced with the help of various supply chains and advanced technology; *patient assumption and hospital discernment* (E10), i.e. all kinds of doubts related to their health must be properly resolved by the hospital so that a good level of trust is developed between patient and doctor; *flexible setups* (E6), i.e. the operational setup and technical assistance should be flexible so that they can be used to handle multiple situations with minimum modifications in the existing structure. Attaining a flexible setup decreases the installation costs of any organisation, as well as increases its efficiency. Adopting all of the concerned parametres in the Indian healthcare sector can be fruitful, as it will increase its efficacy and will make it more responsive. Finally, having identified and encapsulated all of the concerning factors affecting healthcare, these are divided into four quadrants signifying their respective dependence and driving power, as shown in Figure 1.3. The concluding outcome of these factors according to their respective ranks is shown in Table 1.11. The factors considered can be now be employed by various healthcare agencies in order to improve their processes and make these more agile, as highly relevant need.

1.6 CONCLUSION

This chapter outlined the 10 important factors influencing facility management in healthcare organisations through the study of various types of literature, experts' opinions, and research papers. The TISM and MICMAC analysis are used to establish a contextual relationship among these. These factors are up-to-date and play a pivotal part in the betterment of healthcare services. The model developed by the MCDM technique shows a hierarchy of factors describing the levels of influencing power on the healthcare organisation. This study is beneficial for medical colleges, hospitals, healthcare practitioners, nursing homes, *etc*. Today, Indian healthcare is facing problems because it continues with traditional practices when it should adopt agile techniques that make management significantly more effective and efficient. The important factors described under the MICMAC analysis are highlighted in Table 1.1. The factors with the highest driving power are solemnly independent and are of the most importance; thus, they should be given first preference in terms of changing or adopting new norms in any institution, followed by the hierarchical order of other factors as per Table 1.11. The adoption of these factors will help increase the efficacy of any healthcare institute along with developing its long-term sustainability.

REFERENCES

Christopher, M. (2000). The agile supply chain: Competing in volatile markets. *Industrial Marketing Management*, *29*(1), 37–44.

Christopher, M., & Towill, D. (2001). An integrated model for the design of agile supply chains. *International Journal of Physical Distribution & Logistics Management, 31*(4), 235–246.

Gehani, R. R. (1995). Time-based management of technology. *International Journal of Operations & Production Management*, *15*(2), 19–35.

Guimarães, C. M., & de Carvalho, J. C. (2011, January). Outsourcing in the healthcare sector-a state-of-the-art review. In *Supply Chain Forum: An International Journal* (Vol. 12, No. 2, pp. 140–148). Taylor & Francis.

Kidd, P. T. (1995). *Agile manufacturing: Forging new frontiers*. Addison-Wesley Longman Publishing Co., Inc.

Kumar, K., Dhillon, V. S., Singh, P. L., & Sindhwani, R. (2019). Modeling and analysis for barriers in healthcare services by ISM and MICMAC analysis. In M. Kumar, R. Pandey, & R. Kumar (Eds.), *Advances in interdisciplinary engineering*. Lecture Notes in Mechanical Engineering (pp. 501–510). Springer, Singapore. https://doi.org/10.1007/978-981-13-6577-5_47

Kumar, R., Kumar, V., and Singh, S. (2017a). Modeling and analysis on supply chain characteristics using ISM technique. *Apeejay Journal of Management and Technology*, *12*(1 & 2).

Kumar, R., Kumar, V., and Singh, S. (2017b). Work culture enablers: Hierarchical design for effectiveness & efficiency. *International Journal of Lean Enterprise Research (IJLER)*, *2*(3), 189–201.

Mittal, V. K., Sindhwani, R., Shekhar, H., & Singh, P. L. (2019). Fuzzy AHP model for challenges to thermal power plant establishment in India. *International Journal of Operational Research*, *34*(4), 562–581.

Mittal, V. K., Sindhwani, R., Singh, P. L., Kalsariya, V., & Salroo, F. (2018). Evaluating significance of green manufacturing enablers using MOORA method for Indian manufacturing sector. In S. Singh, P. Raj, & S. Tambe (Eds.), *Proceedings of the international conference on modern research in aerospace engineering*. Lecture Notes in Mechanical Engineering (pp. 303–314). Springer, Singapore. https://doi.org/10.1007/978-981-10-5849-3_30

Naylor, J. B., Naim, M. M., & Berry, D. (1999). Lefacility management: Integrating the lean and agile manufacturing paradigms in the total supply chain. *International Journal of Production Economics*, *62*(1–2), 107–118.

Raj, S.A., Sudheer, A., Vinodh, S., & Anand, G. (2013). A mathematical model to evaluate the role of facility management factors and criteria in a manufacturing environment. *International Journal of Production Research*, *51*(19), 5971–5984.

Shanker, K., Shankar, R., & Sindhwani, R. (Eds.). (2019). *Advances in industrial and production engineering*: *Select proceedings of FLAME 2018*. Springer.

Sherehiy, B., Karwowski, W., & Layer, J. K. (2007). A review of enterprise agility: Concepts, frameworks, and attributes. *International Journal of Industrial Ergonomics*, *37*(5), 445-460.

Sindhwani, R., and Malhotra, V. (2017). A framework to enhance agile manufacturing system: A total interpretive structural modelling (TISM) approach. *Benchmarking: An International Journal*, *24*(2), 467–487. https://doi.org/10.1108/BIJ-09-2015-0092

Sindhwani, R., Mittal, V. K., Singh, P. L., Kalsariya, V., & Salroo, F. (2018). Modelling and analysis of energy efficiency drivers by fuzzy ISM and fuzzy MICMAC approach. *International Journal of Productivity and Quality Management*, *25*(2), 225–244.

Sindhwani, R., Singh, P. L., Chopra, R., Sharma, K., Basu, A., Prajapati, D. K., & Malhotra, V. (2019). Facility management evaluation in the rolling industry: A case study. In *Advances in industrial and production engineering* (pp. 753–770). Springer, Singapore.

Sindhwani, R., Singh, P. L., Iqbal, A., Prajapati, D. K., & Mittal, V. K. (2019). Modeling and analysis of factors influencing facility management in healthcare organizations: An

ISM approach. In R. Sindhwani, P. L. Singh, A. Iqbal, D. K. Prajapati, & V. K. Mittal (Eds.), *Advances in industrial and production engineering.* Lecture Notes in Mechanical Engineering (pp. 683–696). Springer, Singapore.

Stelson, P., Hille, J., Eseonu, C., & Doolen, T. (2017). What drives continuous improvement project success in healthcare?*International Journal of Health Care Quality Assurance*, *30*(1), 43-57.

Talib, F., & Rahman, Z. (2015). An interpretive structural modelling for sustainable healthcare quality dimensions in hospital services. *International Journal of Qualitative Research in Services*, *2*(1), 28–46.

Vinodh, S., & Devadasan, S. R. (2011). Twenty criteria based facility management assessment using fuzzy logic approach. *The International Journal of Advanced Manufacturing Technology*, *54*(9–12), 1219–1231.

Vinodh, S., Kumar, V. U., & Girubha, R. J. (2012). Thirty-criteria-based facility management assessment: A case study in an Indian pump manufacturing organisation. *The International Journal of Advanced Manufacturing Technology*, *63*(9–12), 915–929.

Warfield, J. N. (1974). Developing interconnection matrices in structural modeling. *IEEE Transactions on Systems, Man, and Cybernetics, 4*(1), 81–87.

Yusuf, Y. Y., Sarhadi, M., & Gunasekaran, A. (1999). Agile manufacturing: The drivers, concepts and attributes. *International Journal of Production Economics*, *62*(1–2), 33–43.

ANNEXURE I

Sr. No.	Enabler	Comparison of Enabler	Y/N	Relationship of the Enabler with the Other Enabler
E1 Organisational Structure of the Healthcare Industry				
1	E1-E2	The organisational structure of the healthcare industry affects talent management.	Y	The organisation and workforce are inclusive.
2	E2-E1	Talent management affects the organisation structure of the healthcare industry.	N	
3	E1-E3	The organisational structure of the healthcare industry affects conflict management.	Y	Management and employees are integral parts of an organisation.
4	E3-E1	Conflict management affects the organisational structure of the healthcare industry.	N	
5	E1-E4	The organisational structure of the healthcare industry affects inter-connectivity management.	Y	In an organisation, there should be cooperation among different units.
6	E4-E1	Inter-connectivity management affects the organisational structure of the healthcare industry	N	

(*Continued*)

Sr. No.	Enabler	Comparison of Enabler	Y/N	Relationship of the Enabler with the Other Enabler
7	E1-E5	The organisational structure of the healthcare industry affects the enterprise-wide integration of learning.	Y	Good training enhances the competitiveness of the organisation.
8	E5-E1	The enterprise-wide integration of learning affects the organisational structure of the healthcare industry.	N	
9	E1-E6	The organisational structure of the healthcare industry affects customer feedback management.	Y	Units in an organisation must be flexible in terms of technical and operational setup.
10	E6-E1	Customer feedback management affects the organisational structure of the healthcare industry	N	
11	E1-E7	The organisational structure of the healthcare industry affects market awareness.	N	
12	E7-E1	Market awareness affects the enabler organisational structure of the healthcare industry.	N	
13	E1-E8	The organisational structure of the healthcare industry affects external collaboration.	N	
14	E8-E1	External collaboration affects the organisational structure of the healthcare industry.	N	
15	E1-E9	The organisational structure of the healthcare industry affects the adoption of IT tools/technologies.	N	
16	E9-E1	IT tools/technologies adoption affects the organisational structure of the healthcare industry.	N	
17	E1-E10	The organisational structure of the healthcare industry affects the enabler patient assumption and hospital discernment.	N	
18	E10-E1	Patient assumption and hospital discernment affects the enabler organisational structure of the healthcare industry	N	

(*Continued*)

Sr. No.	Enabler	Comparison of Enabler	Y/N	Relationship of the Enabler with the Other Enabler
E2 Talent Management				
1	E2-E3	Talent management affects conflict management.	Y	People with good skills should work in accordance with management.
2	E3-E2	Conflict management affects talent management.	N	
3	E2-E4	Talent management affects inter-connectivity management.	Y	Workers can only be motivated once there is inter- and intra-departmental dependence.
4	E4-E2	Inter-connectivity management affects talent management.	N	
5	E2-E5	Talent management affects the enterprise-wide integration of learning.	Y	Good training ensures skill enhancement.
6	E5-E2	The enterprise-wide integration of learning affects talent management.	N	
7	E2-E6	Talent management affects customer feedback management.	Y	A highly skilled workforce can use the technical and operational setup better.
8	E6-E2	Customer feedback management affects talent management.	N	
9	E2-E7	Talent management affects market awareness.	Y	Being multi-skilled means having some knowledge of market status as well.
10	E7-E2	Market awareness affects talent management.	N	
11	E2-E8	Talent management affects external collaborations.	Y	A flexible workforce is capable of outsourcing products and services.
12	E8-E2	External collaborations affect talent management	N	
13	E2-E9	Talent management affects the adoption of IT tools/technologies.	Y	Skill enhancement involves the adoption of IT technologies.
14	E9-E2	IT tools/technologies adoption affects talent management.	N	
15	E2-E10	Talent management affects the enabler patient assumption and hospital discernment.	Y	Clearing patients' doubts is accomplished by highly skilled people.
16	E10-E2	Patient assumption and hospital discernment affect talent management.	N	

(*Continued*)

Sr. No.	Enabler	Comparison of Enabler	Y/N	Relationship of the Enabler with the Other Enabler
E3 Conflict management				
1	E3-E4	Conflict management affects inter-connectivity management.	Y	Less restriction ensures proper synchronization.
2	E4-E3	Inter-connectivity management affects conflict management.	N	
3	E3-E5	Conflict management affects the enterprise-wide integration of learning.	Y	If there is good training for individuals, then management becomes efficient and easy.
4	E5-E3	The enterprise-wide integration of learning affects conflict management.	N	
5	E3-E6	Conflict management affects customer feedback management.	Y	The coherence of management and employees is ensured by flexible technical and operational setups.
6	E6-E3	Customer feedback management affects conflict management.	N	
7	E3-E7	Conflict management affects market awareness.	N	
8	E7-E3	Market awareness affects conflict management.	N	
9	E3-E8	Conflict management affects external collaborations.	N	
10	E8-E3	External collaborations affect conflict management.	N	
11	E3-E9	Conflict management affects the adoption of IT tools/ technologies.	N	
12	E9-E3	IT tools/technologies adoption affects the enabler conflict management.	N	
13	E3-E10	Conflict management affects patient assumption and hospital discernment.	N	
14	E10-E3	Patient assumption and hospital discernment affect conflict management.	N	
E4 Inter-Connectivity Management				
1	E4-E5	Inter-connectivity management affects the enterprise-wide integration of learning.	N	

(*Continued*)

Sr. No.	Enabler	Comparison of Enabler	Y/N	Relationship of the Enabler with the Other Enabler
2	E5-E4	The enterprise-wide integration of learning affects inter-connectivity management.	N	
3	E4-E6	Inter-connectivity management affects customer feedback management.	Y	Fewer restrictions with a flexible setup prove to be fruitful.
4	E6-E4	Customer feedback management affects inter-connectivity management.	N	
5	E4-E7	Inter-connectivity management affects market awareness.	N	
6	E7-E4	Market awareness affects inter-connectivity management.	N	
7	E4-E8	Inter-connectivity management affects external collaborations.	N	
8	E8-E4	External collaborations affect inter-connectivity management.	N	
9	E4-E9	Inter-connectivity management affects the adoption of IT tools/ technologies.	N	
10	E9-E4	IT tools/technologies adoption affects inter-connectivity management.	Y	Different units can work together with less restriction by using information transfer through IT equipment.
11	E4-E10	Inter-connectivity management affects patient assumption and hospital discernment.	N	
12	E10-E4	Patient assumption and hospital discernment affect inter-connectivity management.	N	
E5 Enterprise-Wide Integration of Learning				
1	E5-E6	The enterprise-wide integration of learning affects customer feedback management.	N	
2	E6-E5	Customer feedback management affects the enterprise-wide integration of learning.	N	
3	E5-E7	The enterprise-wide integration of learning affects market awareness.	N	

(*Continued*)

Sr. No.	Enabler	Comparison of Enabler	Y/N	Relationship of the Enabler with the Other Enabler
4	E7-E5	Market awareness affects the enterprise-wide integration of learning.	N	
5	E5-E8	Enterprise-wide integration of learning affects external collaborations.	N	
6	E8-E5	External collaborations affect the enterprise-wide integration of learning.	N	
7	E5-E9	The enterprise-wide integration of learning affects the adoption of IT tools/technologies.	Y	IT sector knowledge is an essential part of workforce training.
8	E9-E5	IT tools/technologies adoption affects the enterprise-wide integration of learning.	N	
9	E5-E10	The enterprise-wide integration of learning affects patient assumption and hospital discernment.	N	
10	E10-E5	Patient assumption and hospital discernment affect the enterprise-wide integration of learning.	N	
E6 Customer Feedback Management				
1	E6-E7	Customer feedback management affects market awareness.	N	
2	E7-E6	Market awareness affects customer feedback management.	Y	The technical, operational, and infrastructural setup used is cost-effective.
3	E6-E8	Customer feedback management affects external collaborations.	N	
4	E8-E6	External collaborations affect customer feedback management.	N	
5	E6-E9	Customer feedback management affects the adoption of IT tools/technologies.	N	
6	E9-E6	IT tools/technologies adoption affects customer feedback management.	Y	Information Technology ensures operational flexibility
7	E6-E10	Customer feedback management affects patient assumption and hospital discernment.	N	

(Continued)

Sr. No.	Enabler	Comparison of Enabler	Y/N	Relationship of the Enabler with the Other Enabler
8	E10-E6	Patient assumption and hospital discernment affect customer feedback management.	N	
E7 Market Awareness				
1	E7-E8	Market awareness affects external collaborations.	N	
2	E8-E7	External collaborations affect market awareness.	N	
3	E7-E9	Market awareness affects the adoption of IT tools/technologies.	N	
4	E9-E7	IT tools/technologies adoption affects market awareness.	N	
5	E7-E10	Market awareness affects patient assumption and hospital discernment.	Y	Patient diagnoses and prescriptions should be accurate, in accordance with the present operational and technical setup.
6	E10-E7	Patient assumption and hospital discernment affect market awareness.	N	
E8 External Collaborations				
1	E8-E9	External collaborations affect the adoption of IT tools/technologies.	N	
2	E9-E8	IT tools/technologies adoption affects external collaborations.	N	
3	E8-E10	External collaborations affect patient assumption and hospital discernment.	N	
4	E10-E8	Patient assumption and hospital discernment affect external collaborations.	N	
E9 IT Tools/Technologies Adoption				
1	E9-E10	IT tools/technologies adoption affects patient assumption and hospital discernment.	N	
2	E10-E9	Patient assumption and hospital discernment affect the adoption of IT tools/technologies.	N	

2 Assessment of Optimum Automotive Brake Friction Formula by Entropy-VIKOR Technique

Vishal Ahlawat, Parinam Anuradha, Sunil Nain, Sanjay Kajal, Praveen Tewatia, and Mukesh Kumar

CONTENTS

2.1 Introduction....................23
2.2 Formulation of Composition and Testing....................26
2.3 Optimization....................26
2.4 Elementary/Preliminary Structure....................28
2.4.1 Describing PAC, Alternatives, and Forming a Decision Matrix (DM)....................28
2.4.2 Computation of Entropy (h_i)....................29
2.4.3 Determination of Criteria Weight....................30
2.5 Ranking with VIKOR....................30
2.5.1 Utility Measure....................30
2.5.2 Regret Measure....................31
2.5.3 VIKOR Index Ω_j....................31
2.6 Results and Discussion....................32
Conclusions....................39
References....................41

2.1 INTRODUCTION

Multi-criteria decision-making (MCDM) tools and techniques have been employed for solving different multi-criteria decision-making problems that are related to the fields of supply chain, materials selection, sustainable energy, environment, operations research, *etc*. In recent years, researchers have extended the application of MCDM techniques in optimizing brake friction materials that contain more than one conflicting criteria. Brakes are an essential part of a vehicle that is used to retard the vehicle's speed whenever required. A consistent and stable coefficient of friction between the brake friction material and the drum/disc plays an important role in braking (Ahlawat et al., 2018; Bijwe, 1997; Day, 2014). Furthermore, there are some other important characteristics that a brake friction material should possess, i.e. high resistance to wear, minimum heat generation between the rubbing surfaces, effective dissipation of heat, resistance to fade, minimum squeal, and good pedal feel (Puhn, 1985). The brake friction material is composed of several ingredients. Every ingredient, which will be used in composition, has a specific function and should justify the cost with respect to its performance. Variation in the amount of an ingredient may greatly affect frictional performance. Therefore, special attention must be given to the appropriate selection of ingredients and their amount in regards to braking performance.

Brake friction composites may not afford a similar response to all of the braking conditions due to changes in interfacial interactions with the counter surface, availability of ingredients with dissimilar character, and microstructure (Dante, 2015). Modifications in interfacial interactions result in a non-deterministic wear mechanism and generate complexities in predicting the performance. Making the right selection requires high-end expertise in the materials of each category, experience in manufacturing, and testing. Otherwise, it becomes challenging to choose a suitable brake friction material, and trial-and-error remains the only option, which is an expensive and time-consuming task. The selection of suitable brake friction formula on the basis of multiple performance assessment criteria (PAC) can be considered a multi-criteria decision-making (MCDM) problem and may be solved using MCDM techniques because of their ease of implementation, the accuracy of results, and the wide acceptability in the areas of science, management, and engineering (Jahan et al., 2010). There is a significant amount of MCDM techniques available in the literature, which have been applied for solving research problems associated with the real domain containing multiple conflicting criteria. The best suitable brake friction composite out of raw and milled fly ash composites, as well as commercially available brake friction composite specimens, were obtained by combining entropy and VIKOR methods (Ahlawat et al., 2020). Specimens underwent tribo-tests on wear and friction test machines at 3.3 m/s and 5 m/s of sliding velocities under the action of 200 N normal load. The optimum brake friction formula was milled fly ash with 30 wt% content, which performed better at both sliding parametres. A weight determination model, i.e. the Analytic Hierarchy Process (AHP), was used to compute the weight of friction composite encompassing the mixture of Kevlar with lapinus, and lapinus with wollastonite fibers (Singh et al., 2018). Criteria weight was cast off with the VIKOR model of ranking in order to find the best alternative brake friction formula. It was discovered that the use of Kevlar fiber (2.5 wt%) with lapinus fiber (27.5 wt%) offered the best braking performance. Hybridization of the entropy model with grey relational analysis (GRA) was suggested to obtain the best alternate amongst four developed brake friction materials (Kumar et al., 2016). The first two compositions contained 5 and 10 wt% of Kevlar fiber without adding any natural fibers, and the remaining two had natural fibers at 5 wt% and 10 wt% without adding Kevlar fibers. The performance assessment criteria (PAC) for optimization were friction coefficient (FC), fade % and recovery %, wear, cost, and friction fluctuation. Based on the optimization outcomes, the brake friction composite having a 10 wt% of natural fibers produced a greater degree of relationship and was selected as the optimum brake friction composite. Multi-walled carbon nanotubes (MWCNT) and nano clay were used together in different combinations for fabricating brake friction composites and tested for frictional performance (Singh et al., 2015). Preference selection index (PSI) was chosen for optimization based on the experimental values of PAC, i.e. FC, wear rate, stability, variability, fade % and recovery %, fluctuation in FC, and rise in disc temperature. The PSI technique suggested a 2.5 wt% of nano-clay as optimal content and the corresponding composition as an optimum composition for brake friction composite development. The AHP-TOPSIS combinatorial approach was tested for optimizing the brake friction formula amongst the formulated fly ash composite specimens. AHP determined the PAC weights and ranked the available friction composites (Satapathy et al., 2010). Changes in PAC at different tribo-test conditions were predicted relatively well, with a higher degree of accuracy for finding an optimal formulation. Variations in the fabrication process variables influenced the wear and friction characteristics of brake specimens and system productivity. Therefore, the optimization of process variables was done, and the artificial neural network (ANN) approach was employed in order to acquire the best combination of fabrication process variables for high wear resistance (Aleksendrić & Senatore, 2012). Results showed that pressure applied on the die with cycle time, die temperature, and curing time greatly affected the wear of tested specimens. The ANN model and backpropagation algorithm were used to attain optimum brake friction material out of eight friction compositions designed with a baseline friction composition (Daei et al., 2016). The Taguchi approach for experiment design formulated by the ANN was applied in order to deduce the optimum number of test runs. The optimization results showed that the proposed approach proved to be useful in optimization for an actual demonstration.

The availability of multiple numbers of MCDM techniques creates confusion among the researchers on deciding the appropriate one to apply and which one to omit (Kumar et al., 2017a; Sindhwani et al., 2018). However, every problem statement has certain constraints that help the decision-maker in creating a selection of appropriate techniques. Multi-criteria problems require weights of selected performance assessment criteria (PAC) in order to reach an optimum alternative. These weights may be obtained either by subjective or objective approaches. Subjective weighting approaches cover the trade-off method, ranking method, pair-wise comparison, rating method, *etc.*, whereas the objective methods include the entropy model, criteria importance through inter-criteria correlation (CRITIC), variation coefficient approach, TOPSIS, *etc.* (Mittal et al., 2019; Zardari et al., 2015). Every method of solution distinguishes itself in the form of solution ease, applicability, and accuracy. Such weight determination approaches provide the cardinal or ordinal values to the PAC and define their importance relatively. The computed values of criteria weight are amalgamated with MCDM methods for assessing the overall ranking of alternatives (Kumar et al., 2017b; Kumar et al., 2019). The objective approach provides the weight of PAC using certain mathematical models without the interference of designer/decision-maker, whereas subjective methods take designer/decision-maker preference order (Mittal et al., 2018).

In the present investigation, no interference of the decision-maker was involved, and all of the PAC were given equal importance. Thus, the entropy method, an objective method of weight determination, was selected and combined with the VIšekriterijumsko KOmpromisno Rangiranje (VIKOR) approach for ranking. Weight determination includes all of the experimentally assessed values of PAC in order to maintain a relation among the designed brake friction formula (Munier, 2011). This lessens the adverse effect from undesired values and provides more accurate and reasonable ranking calculations. This combinatorial MCDM approach, i.e. entropy-VIKOR, offers a compromised solution near the desired values and away from the worst. Alternatives were taken in the form of nine brake friction composite specimens, which were fabricated with 10 ingredients, including walnut shell powder, graphite, and graphene as friction modifiers. The specimens were tested on wear and friction monitoring machine for FC, CS, CV, friction fluctuation $\Delta\mu$, SWL, and disc-specimen contact temperature at the trailing edge, which were taken as PAC. An optimum brake friction composite out of nine available alternatives based on the six PAC was obtained by applying the entropy-VIKOR multi-criteria decision-making (MCDM) technique.

2.2 FORMULATION OF COMPOSITION AND TESTING

The composition of a brake friction composite usually includes 10–20 ingredients (Ahlawat et al., 2019) in order to provide an adequate friction coefficient, controlled wear rate, effective heat dissipation, pedal feel, *etc*. In the present research, the brake friction composites were fabricated using eight ingredients, including phenol-formaldehyde (PF) as a matrix, $CaCO_3$ as wear the reducing agent, fly ash and rubber as filler materials, Al_2O_3 as the abrasive, banana fiber as reinforcement, and cashew nut shell liquid (CNSL) powder along with varying percentages of walnut shell powder, graphite powder, and graphene powder as friction modifiers. The percentage of walnut shell powder, graphite powder, and graphene powder was varied from 0 to 3 wt%, with CNSL powder which are shown in Table 2.1. Brake friction composites comprised of 1 wt%–3 wt% of walnut shell powder were labelled as WSFC-1, WSFC-2, and WSFC-3 respectively. In a similar manner, graphite and graphene composites were labelled as GTFC-1, GTFC-2, GTFC-3 and GNFC-1, GNFC-2, GNFC-3 respectively. Composite specimens were fabricated using a compression moulding machine, as per the detailed composition presented in Table 2.1. Figure 2.1 shows the process flow of specimen development with operating conditions.

2.3 OPTIMIZATION

In the present study, the optimum brake friction formula out of nine developed friction formulations containing more than one conflicting criteria was obtained by applying the entropy-VIKOR multi-criteria approach. The developed specimens were tribo-tested at 3.3 m/s and 9.9 m/s under the action of 50 N and 150 N normal loads. The values obtained from experimentation were used in the entropy model to find the weights of the performance assessment criteria. These weights establish a relationship among the designed friction composites, which were cast off further in the VIKOR technique to rank the brake friction composites. The entropy technique

TABLE 2.1
Detailed Composition and Designation

Labelled Specimens	Ingredients (%)								
	PF Resin	$CaCO_3$	Fly Ash	Al_2O_3	Banana Fiber	Rubber	CNSL	Walnut Shell Powder	Graphite	Graphene
WSFC-1	15	30	20	5	5	4	20	1	0	0
WSFC-2	15	30	20	5	5	4	19	2	0	0
WSFC-3	15	30	20	5	5	4	18	3	0	0
GTFC-1	15	30	20	5	5	4	20	0	1	0
GTFC-2	15	30	20	5	5	4	19	0	2	0
GTFC-3	15	30	20	5	5	4	18	0	3	0
GNFC-1	15	30	20	5	5	4	20	0	0	1
GNFC-2	15	30	20	5	5	4	19	0	0	2
GNFC-3	15	30	20	5	5	4	18	0	0	3

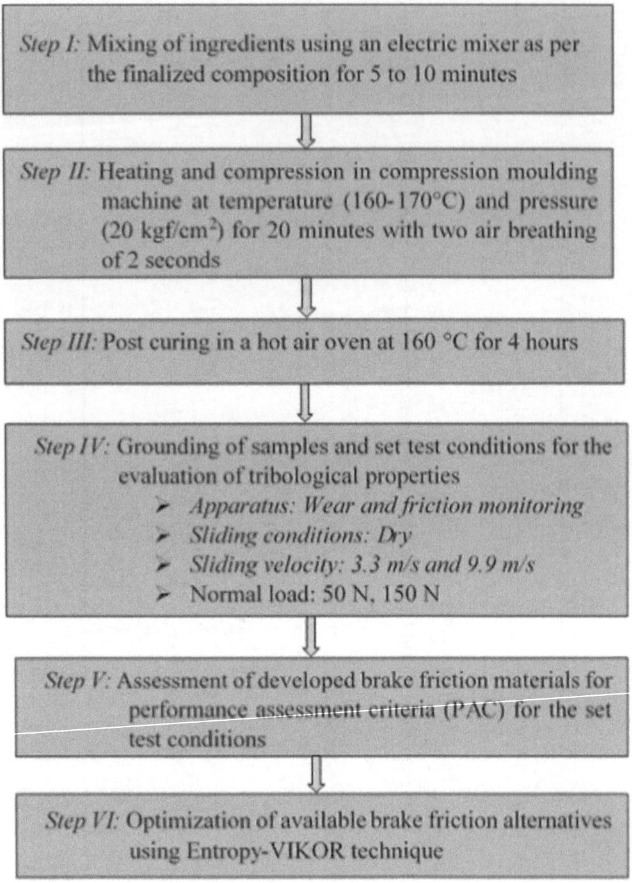

FIGURE 2.1 Process Flow from Specimen Development to Optimization.

consisted of two stages: The first stage includes the criteria and alternatives definition and the formation of decision matrix (DM), while the second stage involves the computation of criteria weights. In the next stage, the VIKOR technique was used to rank the available alternatives by using these weights. The detailed process flow of the entropy-VIKOR technique is shown in Figure 2.2.

2.4 ELEMENTARY/PRELIMINARY STRUCTURE

2.4.1 Describing PAC, Alternatives, and Forming a Decision Matrix (DM)

Nine friction composites were fabricated with varying amounts of 1, 2, and 3 wt % of walnut shell powder, graphite powder, and graphene powder and were accordingly chosen as alternatives. These specimens were tested for FC, CS, CV, $\Delta\mu$, SWL, and the disc-specimen contact temperature at the trailing edge, which were considered as performance assessment criteria (PAC). PAC and alternatives are described in

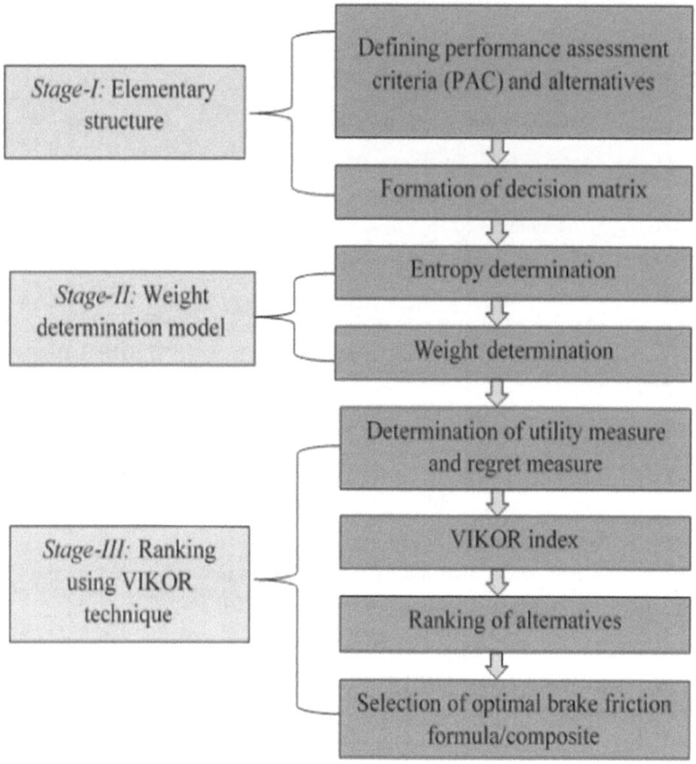

FIGURE 2.2 Process Flow Showing the Entropy-VIKOR Model.

Table 2.2. The *DM* is constructed using M number of criteria, i.e. 6, and N number of alternatives, i.e. 9, of an order M × N.

$$DM_{M\times N} = \begin{bmatrix} a_{11} & a_{12} & \dots & a_{1N} \\ a_{21} & a_{22} & \dots & a_{2N} \\ \vdots & \vdots & \dots & \vdots \\ a_{M1} & a_{M2} & \dots & a_{MN} \end{bmatrix} \tag{2.1}$$

Where, a_{ij} = experimentally obtained value corresponds to ith criteria of jth alternate; i = 1, 2, …, *M*; j = 1, 2, …, *N*.

2.4.2 Computation of Entropy (H_I)

Entropy h_i is a measure of diversity in the values of criteria obtained from experimentation and can be computed using the following equation:

$$h_i = -\left(\sum_{j=1}^{N} p_{ij} \ln(p_{ij}) \right) \Big/ \ln(N) \tag{2.2}$$

TABLE 2.2
Description of PAC and Alternatives

PAC (M)	Definition and Level of Requirement	Alternatives (N)
PAC-1: Average friction coefficient (FC)	Average FC is an average of FC values obtained after each test; a higher value is recommended.	N1: WSFC-1
		N2: WSFC-2
		N3: WSFC-3
PAC-2: Coefficient of stability (CS)	CS = $(\mu_{average})/(\mu_{max})$; a higher value is recommended.	N4: GTFC-1
		N5: GTFC-2
PAC-3: Coefficient of variability (CV)	CV = $(\mu_{min})/(\mu_{max})$; a higher value is recommended.	N6: GTFC-3
		N7: GNFC-1
		N8: GNFC-2
PAC-4: Friction fluctuation ($\Delta\mu$)	$(\Delta\mu) = (\mu_{max}) - (\mu_{min})$; a lower value is recommended.	N9: GNFC-3
PAC-5: Specific wear loss (SWL)	$SWL = \frac{W_1 - W_2}{\rho\ d\ P_n} \times 10^{-8},\ \frac{cm^3}{N-m}$; a lower value is recommended.	
PAC-6: Temperature rise	Temperature rises at the contact of disc and specimen; a lower value is recommended.	

Here, μ = FC, $\mu_{average}$ = average FC, μ_{min} = minimum FC; μ_{max} = maximum FC; ρ = composite actual density (g/cm^3), d = track diameter of wear (mm), P_n = normal load acting (N); W_2 = weight after conducting experiment (g); W_1 = weight before conducting experiment (g)

Where, p_{ij} are the normalized values of criteria for the available alternatives. The normalized values are required to transform the different units of criteria into a single measurable unit and can be computed as: $p_{ij} = \frac{a_{ij}}{\sum_{j=1}^{N} a_{ij}}$; $i = 1, 2, \ldots, M$ and $j = 1, 2, \ldots, N$

2.4.3 Determination of Criteria Weight

Criteria weight can help in establishing a relationship among the designed friction formula. It can be computed by the following equation:

$$\omega_i = \frac{1 - h_i}{\sum_{i=1}^{M} (1 - h_i)} = \frac{d_i}{\sum_{s=1}^{M} d_s} \tag{2.3}$$

2.5 RANKING WITH VIKOR

2.5.1 Utility Measure

Utility measure α_j can be determined as:

$$\alpha_j = \sum_{i=1}^{M} \frac{\omega_i\,[(a_{ij})_{\text{maximum}} - a_{ij}]}{[(a_{ij})_{\text{maximum}} - (a_{ij})_{\text{minimum}}]},\ \text{if i is a benefit criteria} \tag{2.4}$$

$$\alpha_j = \sum_{i=1}^{M} \frac{\omega_i\,[a_{ij} - (a_{ij})_{\text{minimum}}]}{[(a_{ij})_{\text{maximum}} - (a_{ij})_{\text{minimum}}]},\ \text{if i is a cost criteria} \tag{2.5}$$

2.5.2 Regret Measure

Regret measure, as notated by β_j, can be calculated as:

$$\beta_j = Max^a\, of \left\{ \frac{\omega_i\,[(a_{ij})_{\text{maximum}} - a_{ij}]}{[(a_{ij})_{\text{maximum}} - (a_{ij})_{\text{minimum}}]} \right\};\ \text{i} = 1,\ 2,\ \ldots,\ \text{M} \tag{2.6}$$

Here, benefit $(a_{ij})_{\text{maximum}}$ and cost $(a_{ij})_{\text{minimum}}$ criteria can be obtained as:

$$(a_{ij})_{\text{maximum}} = {}^{\text{maximum}}{}_j a_{ij} = \text{maximum}[a_{ij}, j = 1, 2, \ldots, N]$$

$$(a_{ij})_{\text{minimum}} = {}^{\text{minimum}}{}_j a_{ij} = \text{minimum}[a_{ij}, j = 1, 2, \ldots, N]$$

2.5.3 VIKOR Index Ω_j

The VIKOR Index can be computed by the given relationship:

$$\Omega_j = \xi \left(\frac{(\alpha_j - \alpha_j^-)}{(\alpha_j^+ - \alpha_j^-)} \right) + (1 - \xi) \left(\frac{(\beta_j - \beta_j^-)}{(\beta_j^+ - \beta_j^-)} \right) \tag{2.7}$$

Where,

$$\alpha_j^+ = {}^{\text{maximum}}{}_j \alpha_j = \text{maximum}[\alpha_j, j = 1, 2, \ldots, N];$$

$$\alpha_j^- = {}^{\text{minimum}}{}_j \alpha_j = \text{minimum}[\alpha_j, j = 1, 2, \ldots, N];$$

$$\beta_j^+ = {}^{\text{maximum}}{}_j \beta_j = \text{maximum}[\beta_j, j = 1, 2, \ldots, N];$$

$$\beta_j^- = {}^{\text{minimum}}{}_j \beta_j = \text{minimum}[\beta_j, j = 1, 2, \ldots, N];$$

$\xi = 0.5$ and $(1 - \xi)$ = individual regret weight.

The minimum value of the index (Ω_j) can be granted the first rank, which shows the optimum brake friction alternative. Subsequently, the remaining alternatives can be ranked in ascending order with an increase in the value of the VIKOR index. The friction formulation corresponding to rank one brake friction composite can be chosen as the optimum one that will satisfy all the PAC.

2.6 RESULTS AND DISCUSSION

The criteria values of experimentally tested friction composites at 3.3 m/s and 9.9 m/s under normal loads of 50 N and 150 N were used in entropy-VIKOR in order to rank the friction composites. The developed specimens were considered as alternatives, and the tribo-performance parametres FC, CS, CV, $\boldsymbol{\Delta\mu}$, SWL, and the disc-specimen contact temperature at the trailing edge were treated as performance assessment criteria (PAC).

CASE STUDY-1 RANKING AT A 3.3 M/S SLIDING VELOCITY AND 50 N NORMAL LOAD

Six PAC and nine brake friction composites form a decision matrix (*DM*) of an order of (6 × 9) using experimentally assessed values of performance criteria for each alternative and are presented in Table 2.3. The experimentally attained values were normalized from all of the units in order to attain equal significance and to avoid anomalies. Normalized values (p_{ij}) are also called projection values and can be tabulated in the form of the projection matrix, as illustrated in Table 2.4. The next level is obtaining entropy (h_i), which gives us information about diversity in the PAC values of alternatives. This is computed using equation (2) and subsequently used to find the weight of criteria. *E*quation (3) was used to calculate the weight, as tabulated in Table 2.5. The observed values of criteria from experiments are neither good nor bad; however, these affect the tribo-performance. Desired values can be considered as benefit criteria and non-desired as cost criteria. Values of benefit and cost criteria are presented in Table 2.6.

Equations 2.4 and 2.5 give the utility measures, and Equation 2.6 shows the regret measure, which were used to find the VIKOR index from Equation 2.7. Table 2.7 presents the computed value of α_j, β_j, and Ω_j. Table 2.7 presents the ranking of tested brake friction composites based on the VIKOR index. Rank one was ascribed to WSFC-2, as its VIKOR index is lesser. The brake friction composites can be arranged in ascending order from 1 to 9 as WSFC-2>GNFC-1>GTFC-1>WSFC-1>GTFC-2>GTFC-3>GNFC-2>GNFC-3>WSFC-3.

TABLE 2.3
Formation of DM

Criteria	Alternatives								
	WSFC-1	WSFC-2	WSFC-3	GTFC-1	GTFC-2	GTFC-3	GNFC-1	GNFC-2	GNFC-3
PAC-1	0.56	0.5	0.52	0.4	0.43	0.48	0.53	0.48	0.43
PAC-2	0.92	0.90	0.83	0.93	0.87	0.92	0.95	0.90	0.84
PAC-3	0.72	0.92	0.33	0.70	0.75	0.70	0.71	0.48	0.54
PAC-4	0.170	0.04	0.42	0.13	0.12	0.16	0.16	0.28	0.23
PAC-5	1.145	0.626	0.631	1.081	1.403	1.418	0.596	0.686	2.225
PAC-6	43.6	51.2	42.4	44.6	46.1	43.8	42.6	42.2	42.9

TABLE 2.4
Normalization/Projection Matrix

Criteria	Alternatives								
	WSFC-1	WSFC-2	WSFC-3	GTFC-1	GTFC-2	GTFC-3	GNFC-1	GNFC-2	GNFC-3
PAC-1	0.1293	0.1155	0.1201	0.0924	0.0993	0.1109	0.1224	0.1109	0.0993
PAC-2	0.1141	0.1117	0.1030	0.1154	0.1079	0.1141	0.1179	0.1117	0.1042
PAC-3	0.1231	0.1573	0.0564	0.1197	0.1283	0.1194	0.1214	0.0821	0.0924
PAC-4	0.0994	0.0234	0.2456	0.0760	0.0702	0.0936	0.0936	0.1637	0.1345
PAC-5	0.1167	0.0639	0.0643	0.1102	0.1430	0.1445	0.0607	0.0699	0.2268
PAC-6	0.1092	0.1282	0.1062	0.1117	0.1154	0.1097	0.1067	0.1057	0.1074

TABLE 2.5
Computed Values of h_i, d_i, and ω_i

	PAC-1	PAC-2	PAC-3	PAC-4	PAC-5	PAC-6
h_j	0.9973	0.9993	0.9845	0.9346	0.9538	0.9989
d_i	0.0027	0.0007	0.0155	0.0654	0.0462	0.0011
ω_i	0.0206	0.0052	0.1180	0.4969	0.3512	0.0081

TABLE 2.6
Benefit Criteria and Cost Criteria

PAC	(a_{ij})maximum	(a_{ij})minimum
PAC-1	0.56	0.4
PAC-2	0.95	0.83
PAC-3	0.92	0.33
PAC-4	0.04	0.42
PAC-5	0.596	2.225
PAC-6	42.2	51.2

CASE STUDY-2 RANKING AT A 9.9 M/S SLIDING VELOCITY AND 150 N NORMAL LOAD

All of the specimens were tested at a 9.9 m/s sliding velocity and 150 N normal load, and the values obtained for each alternative PAC were used to form the *DM*. Six PAC and nine alternatives make a decision matrix (M × N) of order (6 × 9), where M are the PAC and N are the alternatives. Table 2.8 presents the decision matrix, and Table 2.9 depicts the normalization matrix. Entropy (h_i), which gives us information about the diverse response of PAC, was computed from Equation 2.2 and used subsequently to find the weight of criteria. Equation 2.3 was used to calculate the weight, which is accordingly tabulated in Table 2.10.

The desired values of criteria were chosen as benefit criteria, and undesired values as cost criteria, as tabulated below in Table 2.11. These values were used in computing the values of α_j, β_j, and Ω_j, which are illustrated in Table 2.12.

As observed in Table 2.12, the VIKOR index for the WSFC-2 specimen is the least and is named rank one among all of the tested friction composites. The WSFC-2 specimen resulted as the best alternative at both the moderate and harsh sliding conditions. Ingredients with content related to the specimen can be considered optimum. The ranking of the rest of the alternatives can be

TABLE 2.7
Calculated Values of α_j, β_j, and Ω_j

	WSFC-1	WSFC-2	WSFC-3	GTFC-1	GTFC-2	GTFC-3	GNFC-1	GNFC-2	GNFC-3
α_j	0.3310	0.0245	0.6330	0.2899	0.3363	0.3915	0.2031	0.4337	0.6978
β_j	0.1700	0.0081	0.4969	0.1177	0.1740	0.1771	0.1569	0.3138	0.3511
Ω_j	0.3935	0.0001	0.9524	0.3095	0.4015	0.4459	0.2851	0.6170	0.8515
Ranking	4	1	9	3	5	6	2	7	8

TABLE 2.8
Construction of DM

Criteria	Alternatives								
	WSFC-1	WSFC-2	WSFC-3	GTFC-1	GTFC-2	GTFC-3	GNFC-1	GNFC-2	GNFC-3
PAC-1	0.27	0.24	0.33	0.28	0.27	0.32	0.25	0.27	0.25
PAC-2	0.81	0.69	0.77	0.78	0.78	0.67	0.82	0.78	0.66
PAC-3	0.67	0.94	0.91	0.75	0.86	0.79	0.90	0.86	0.79
PAC-4	0.11	0.02	0.04	0.09	0.05	0.1	0.03	0.05	0.08
PAC-5	0.827	0.905	1.261	0.721	1.013	0.945	0.728	0.838	0.989
PAC-6	78.6	75.1	86.5	75.5	74.3	76.3	76.8	66.8	75.9

TABLE 2.9
Normalization/Projection Matrix

Criteria	Alternatives								
	WSFC-1	WSFC-2	WSFC-3	GTFC-1	GTFC-2	GTFC-3	GNFC-1	GNFC-2	GNFC-3
PAC-1	0.1089	0.0968	0.1331	0.1129	0.1089	0.1290	0.1008	0.1089	0.1008
PAC-2	0.1200	0.1026	0.1134	0.1158	0.1153	0.0997	0.1212	0.1153	0.0968
PAC-3	0.0894	0.1264	0.1216	0.1006	0.1149	0.1056	0.1207	0.1149	0.1059
PAC-4	0.1930	0.0351	0.0702	0.1579	0.0877	0.1754	0.0526	0.0877	0.1404
PAC-5	0.1005	0.1100	0.1533	0.0876	0.1232	0.1149	0.0885	0.1019	0.1202
PAC-6	0.1146	0.1095	0.1261	0.1101	0.1083	0.1113	0.1120	0.0974	0.1107

TABLE 2.10
Computed Values of h_i, d_i, and ω_i

	PAC-1	PAC-2	PAC-3	PAC-4	PAC-5	PAC-6
h_i	0.9973	0.9984	0.9975	0.9445	0.9933	0.9988
d_i	0.0027	0.0016	0.0025	0.0555	0.0067	0.0012
ω_i	0.0390	0.0230	0.0361	0.7898	0.0959	0.0165

TABLE 2.11
Benefit Criteria Values and Cost Criteria Values

PAC	(a_{ij})maximum	(a_{ij})minimum
PAC-1	0.33	0.24
PAC-2	0.82	0.66
PAC-3	0.94	0.67
PAC-4	0.02	0.11
PAC-5	0.72	1.26
PAC-6	78.6	86.5

done in ascending order with a corresponding increase in VIKOR index as follows: WSFC-2>GNFC-1>WSFC-3>GNFC-2>GTFC-2>GNFC-3>GTFC-1>GTFC-3>WSSFC-1.

CONCLUSIONS

The present study is about the assessment of ranking using the entropy-VIKOR optimization technique. The performance assessment criteria (PAC) of fabricated composite specimens were evaluated at moderate and harsh test conditions. Based on evaluation and ranking, the conclusions drawn are as follows:

1. WSFC-2 composite specimens having a 2 wt % of walnut shell powder showed better performance at both the test conditions. Therefore, it is concluded that a 2 wt % concentration of walnut shell powder corresponding to a WSFC-2 friction composite is optimal content and may be proposed for the development of brake linings.
2. The composites specimens in ascending order from rank one to rank nine under the load of 50 N at 3.3 m/s were discovered to be WSFC-2>GNFC-1>GTFC-1>WSFC-1>GTFC-2>GTFC-3>GNFC-2>GNFC-3>WSFC-3 and at a sliding velocity of 9.9 m/s and a normal load of 150 N, the ranking order was WSFC-2>GNFC-1>WSFC-3>GNFC-2>GTFC-2>GNFC-3>GTFC-1>GTFC-3>WSFC-1.

TABLE 2.12
Calculated Values of α_j, β_j, and Ω_j

	WSFC-1	WSFC-2	WSFC-3	GTFC-1	GTFC-2	GTFC-3	GNFC-1	GNFC-2	GNFC-3
α_j	0.8725	0.0823	0.3000	0.6603	0.3492	0.7830	0.1255	0.3025	0.6472
β_j	0.7898	0.0390	0.1755	0.6143	0.2633	0.7021	0.0878	0.2633	0.5265
Ω_j	0.9997	0.0001	0.2286	0.7486	0.3182	0.8846	0.0599	0.2887	0.6819
Ranking	9	1	3	7	5	8	2	4	6

3. A combination of entropy and VIKOR techniques resulted in a compromised, albeit effective and accurate, solution to the multi-criteria problem related to the automotive sector.

The combination of manufacturing process parametres such as mixing time, curing time, and post-curing time may also affect the mechanical and tribo-performance of brake friction composites; hence, these may be required to be optimized. Thus, the entropy-VIKOR multi-criteria approach may be employed to optimize the manufacturing processes parametres of brake friction composites. The entropy-VIKOR approach can be adopted for problems with multiple criteria, and this does not involve any interruption from the designer side.

REFERENCES

Ahlawat, V., Kajal, S., & Parinam, A. (2018). Effect of mechanical milling of fly ash on friction and wear response of brake friction composites. *International Journal of Surface Science and Engineering*, *12*(5/6), 433. https://doi.org/10.1504/IJSURFSE.2018.096754

Ahlawat, V., Kajal, S., & Anuradha, P. (2019). Tribo-performance assessment of milled fly ash brake friction composites. *Polymer Composites, 41*(2), 707–718.

Ahlawat, V., Anuradha, P., & Kajal, S. (2020). Preference selection of brake friction composite using entropy-VIKOR technique. *Materials Today: Proceedings*. ISSN 2214-7853, https://doi.org/10.1016/j.matpr.2020.04.256. (https://www.sciencedirect.com/science/

Aleksendrić, D., & Senatore, A. (2012). Optimization of manufacturing process effects on brake friction material wear. *Journal of Composite Materials*, *46*(22), 2777–2791.

Bijwe, J. (1997). Composites as friction materials: Recent developments in non-asbestos fiber reinforced friction materials—A review. *Polymer Composites*, *18*(3), 378–396.

Daei, A. R., Davoudzadeh, N., & Filip, P. (2016). Optimization of brake friction materials using mathematical methods and testing. *SAE International Journal of Materials and Manufacturing*, *9*(1), 118–122.

Dante, R. C. (2015). *Handbook of friction materials and their applications*. Woodhead Publishing.

Day, A. J. (2014). *Braking of road vehicles*. Butterworth-Heinemann.

Jahan, A., Ismail, M. Y., Sapuan, S. M., & Mustapha, F. (2010). Material screening and choosing methods–A review. *Materials & Design*, *31*(2), 696–705.

Kumar, N., Singh, T., Rajoria, R. S., & Patnaik, A. (2016). Optimum design of brake friction material using hybrid entropy-GRA approach. *MATEC Web of Conferences*, *57*, 03002.

Kumar, K., Dhillon, V. S., Singh, P. L., & Sindhwani, R. (2019). Modeling and analysis for barriers in healthcare services by ISM and MICMAC analysis. In M. Kumar, R. Pandey, & V. Kumar (Eds.), *Advances in interdisciplinary engineering*. Lecture Notes in Mechanical Engineering (pp. 501–510). Springer, Singapore. https://doi.org/10.1007/978-981-13-6577-5_47

Kumar, R., Kumar, V., and Singh, S. (2017a). Modeling and analysis on supply chain characteristics using ISM technique. *Apeejay Journal of Management and Technology*, *12*(1 & 2), 21–30.

Kumar, R., Kumar, V., and Singh, S. (2017b). Modeling and analyzing the impact of lean principles on organizational performance using ISM approach. *Journal of Project Management*, *2*, 37–50.

Mittal, V. K., Sindhwani, R., Shekhar, H., & Singh, P. L. (2019). Fuzzy AHP model for challenges to thermal power plant establishment in India. *International Journal of Operational Research*, *34*(4), 562–581.

Mittal, V. K., Sindhwani, R., Singh, P. L., Kalsariya, V., & Salroo, F. (2018). Evaluating significance of green manufacturing enablers using MOORA method for Indian manufacturing sector. In *Proceedings of the international conference on modern research in aerospace engineering*. Lecture Notes in Mechanical Engineering (pp. 303–314). Springer, Singapore.

Munier, N. (2011). *A strategy for using multicriteria analysis in decision-making: A guide for simple and complex environmental projects*. Springer Science & Business Media.

Puhn, F. (1985). *Brake handbook*. HP Books.

Satapathy, B. K., Majumdar, A., & Tomar, B. S. (2010). Optimal design of flyash filled composite friction materials using combined analytical hierarchy process and technique for order preference by similarity to ideal solutions approach. *Materials & Design*, *31*(4), 1937–1944.

Sindhwani, R., Mittal, V. K., Singh, P. L., Kalsariya, V., & Salroo, F. (2018). Modelling and analysis of energy efficiency drivers by fuzzy ISM and fuzzy MICMAC approach. *International Journal of Productivity and Quality Management*, *25*(2), 225–244.

Singh, T., Patnaik, A., Chauhan, R., & Chauhan, P. (2018). Selection of brake friction materials using hybrid analytical hierarchy process and vise Kriterijumska Optimizacija Kompromisno Resenje approach. *Polymer Composites*, *39*(5), 1655–1662.

Singh, T., Patnaik, A., Gangil, B., & Chauhan, R. (2015). Optimization of tribo-performance of brake friction materials: Effect of nano filler. *Wear*, *324*, 10–16.

Zardari, N. H., Ahmed, K., Shirazi, S. M., & Yusop, Z. B. (2015). *Weighting methods and their effects on multi-criteria decision making model outcomes in water resources management*. Springer.

3 Ranking of Phase Change Materials for Medium Temperature Thermal Energy Accumulation System Using Shannon Entropy, TOPSIS, and VIKOR Methods

Ankit Yadav, Vikas, and Sushant Samir

CONTENTS

3.1 Introduction and Background 44
3.2 Materials and Methods 46
3.2.1 Definition 46
3.2.2 Multi-Criteria Decision-Making Methods 47
3.2.2.1 Criteria Weighting 47
3.2.2.2 TOPSIS Method 48
3.2.2.3 VIKOR Method 48
3.3 Findings and Discussion 49
3.3.1 Criteria Weighting 49
3.3.2 TOPSIS Method 50
3.3.3 VIKOR Method 51
3.4 Conclusion and Suggestions 52
References 53
Annexures A-1: Thermo-Physical Properties of the Candidate Organic PCM 56
Annexures A-2: Ranking of Candidate Organic PCM Based on the TOPSIS Method 58
Annexures A-3: Ranking of Candidate Organic PCM Based on the VIKOR Method 61

3.1 INTRODUCTION AND BACKGROUND

The design of Latent Heat Thermal Energy Accumulation Systems (LHTEAS) entirely depends on the selection of a suitable PCM. PCM are the materials that reversibly change their physical state from solid to liquid during melting or liquid to vapor during vaporization within the phase transition temperature range. During the phase change, such materials accumulate or release a large chunk of thermal energy (Rathod & Kanzaria, 2011; Singh et al., 2019). Due to the rise in energy demand and exploitation of conventional energy resources, LHTEAS is gaining the attention of energy researchers, which in turn necessitates emphasizing the judicious use of PCM for the optimum design of LHTEAS. LHTEAS are being incorporated in various applications, like solar thermal systems (e.g. solar heat pumps, solar domestic water heaters, solar dryers, solar desalination, *etc.*), building thermal management, design of battery packs for optimum thermal performance, and passive or active comfort heating/cooling due to their property of higher energy accumulation density than that of sensible energy accumulation systems (Bilardo et al., 2020; Sardari et al., 2019; Weng et al., 2019). The use of PCM for energy accumulation plays a vital role in reducing thermal waste, curbing environmental pollution, and promoting the sustainable use of available energy resources. However, it is a laborious task to select suitable PCM due to the presence of a large number of commercially available PCM with different thermo-physical properties (Wei et al., 2018). Moreover, it is always advisable to obtain optimum thermal performance at the lowest possible cost. The first and most important property of PCM is the phase change temperature suited to an application of a particular temperature range.

Generally, PCM incorporated in LHTEAS are single material, metals and alloys, or a eutectic mixture of two or more substances. From the literature, it is discovered that a single PCM is not capable of fulfilling all of the desired properties for an LH TEAS (Zayed et al., 2019). Therefore, the selection of PCM is not an easy task, as it is comprised of many variables and factors. The PCM are selected based mostly upon one's prior knowledge in terms of designing such systems or by a trial-and-error method. However, such a strategy of selection does not guarantee consistent success in the selection of PCM. Hence, a reliable and logical approach should be used by material experts in sorting and selecting PCM. The multi-parametre nature of the design process of LHTEAS and diverse classes of PCM create a scope for the development of a systematic approach.

Every PCM has distinct characteristics, and a researcher has to search and select from the available pool of alternatives according to one or more properties or based on prior experience. There is no standard procedure in selecting an optimal PCM while considering all the properties. Accordigly, this dependency on more properties makes this problem a multi-attribute decision-making (MADM) problem. Previously, the MADM approach has successfully been used in various fields or areas, such as manufacturing processes (Venkata Rao, 2009), financial decisions, composite (Milani et al., 2011), or other material selection (Athawale & Chakraborty, 2012; Yazdani & FarokhPayam, 2015). History shows that the decision-making process, like multi-criteria decision making (MCDM,) plays a vital role in the selection of materials. MCDM has sub-branches, such as multi-objective decision-making (MODM) and

MADM. Using these techniques, researchers or experts can select optimum material considering multiple criteria instead of single criterion (Mittal et al., 2019; Shanker et al., 2019; Sindhwani et al., 2018; Sindhwani & Malhotra, 2017; Sindhwani, Singh, Chopra, et al., 2019; Sindhwani, Singh, Iqbal, et al., 2019). Therefore, the use of these techniques is constructive for selection or decision-making processes. Various researchers have used these techniques either in the conventional or hybrid sense or with certain modifications. Other methods, like Ashby, are also available, but if the number of parametres is more, then the Ashby method will not be able to rank the alternatives. As a result, MADM is employed as a second branch of MCDM. Moreover, the TOPSIS and VIKOR methods can handle a wide range of selection problems for material characterization. Various comparative studies have been done using the TOPSIS, VIKOR, and some other MCDM methods by Chakraborty & Chatterjee, 2013; Liu et al., 2014; Yazdani & FarokhPayam, 2015. Multiple studies have analyzed the suitability and consistency of MADM methods.

In recent years, MADM techniques are getting more attention in the material selection field. Çolak & Kaya, 2020 evaluated various energy storage alternatives, such as electrical, thermal, compressed air, chemical, *etc.* using MCDM. Singh et al., 2019 used MADM and MODM methods for ranking and selection of nanoparticle-enhanced PCM used in designing a conical LHTEAS. Graphene was ranked as the best nanoparticle candidate in enhancing the conducting properties of PCM, followed by aluminium and alumina. Rastogi et al., 2015 presented the MCDM approach for ranking and making the selection of PCM for domestic water or space heating, ventilation, and air-conditioning application using two figures of merits (FOM) to grade performance. Saaty, 1990 presented a material selection procedure based on a combination of AHP. However, very few studies are available regarding the selection of optimum PCM using the MADM approach. Various researchers also used the TOPSIS and Fuzzy TOPSIS for PCM selection. Xu et al., 2017 used MCDM to select suitable PCM for application in the refrigeration system and found it a suitable technique for the selection procedure. The results show that if precise values of performance ratings are available, then TOPSIS is suitable; otherwise, for imprecise values, Fuzzy TOPSIS is more appropriate to apply. Beltrán & Martínez-Gómez, 2019 used COPRAS-G, TOPSIS, VIKOR, ARAS, and building energy simulation methods to select applicable PCM candidates for building wallboard composite. The results concluded that the building energy simulation along with consideration of environmental factors are necessary to contrast the results obtained from various MCDM selection methods. Wei et al., 2018 presented the review of PCM selection principles for high-temperature PCM. VIKOR is also an MADM approach, but only limited literature shows Shannon Entropy (Hafezalkotob & Hafezalkotob, 2016), TOPSIS, and VIKOR in selecting optimal PCM (Pereira da Cunha & Eames, 2016). Oluah et al., 2020 used the entropy method to determine the weights and the TOPSIS method for ranking 11 PCM in building Trombe-wall. The thermal conductivity of PCM was ranked the highest weight for selection.

The present study is carried out in the absence of investigations in the literature on PCM selection techniques for medium temperature LHTEAS applications. The selection of optimal PCM for LHTEAS will improve system performance and reliability. Based on the feasibility analysis presented by the German Technical Cooperation (GTZ), GIZ, and GoI MNRE, 2011, the temperature range of organic PCM for LHTEAS is selected between 40–120 °C. In this chapter, the organic PCM alternatives available for a given temperature range were chosen from the literature/database. The selected alternatives were further analyzed and ranked using the Shannon Entropy, TOPSIS, and VIKOR methods. The PCM were listed based on six criteria, i.e. melting temperature, thermal conductivity, mass density, specific heat capacity, latent heat, and kinematic viscosity.

3.2 MATERIALS AND METHODS

3.2.1 Definition

The desired specific range of working temperature in thermal systems can be achieved with the help of PCM due to the exchange of latent heat of phase change. Therefore, the phase transition temperature of the PCM should lie within the desired temperature range, which in turn leads to a more stable and nearly isothermal latent energy storage system. Along with the selected temperature range, other properties of the PCM are of great importance as well. Based on the literature, the following features of the PCM are essential:

- High specific heat, the high heat of fusion per unit volume, and high thermal conductivity in order to make the LHTEAS more compact and efficient.
- Reversible phase change.
- Phase transition temperature (solidus and liquidus temperature) lies well within the desired application range.
- Little to no volumetric change during the phase transition.
- No or insignificant sub-cooling and super-cooling.
- Suitability with different materials of encapsulation and framework.
- Eco-friendly, i.e. chemically stable, harmless, non-toxic, *etc.*
- Ample availability.

Accounting for all of the desirable properties and in order to meet all of the abovementioned design requirements, the following thermophysical properties are considered for analysis:

- The melting temperature (T_m)
- Thermal conductivity (k)
- Density (ρ)
- Specific heat (C_p)
- Latent heat (L)
- Kinematic viscosity (ν)

The list of the organic PCM selected for this study is taken from the literature/database (Vitorino et al., 2016), which is presented in Annexure A-1.

3.2.2 Multi-Criteria Decision-Making Methods

A decision matrix was formulated, and the weight of each property (i.e. criterion) was calculated using the Shannon Entropy approach. The calculated weights of properties were further used to prioritize and rank the PCM with the help of the TOPSIS and VIKOR methods. The mathematical expressions of Shannon Entropy, TOPSIS, and VIKOR method are presented in the subsequent sections.

3.2.2.1 Criteria Weighting

Shannon Entropy is a probabilistic approach used to measure uncertainty in the given data. According to this method, the broad distribution of provided data is an indication of higher uncertainty (Beltrán & Martínez-Gómez, 2019; Oluah et al., 2020; Zou et al., 2006). The name has its roots in the study of entropy generation of thermodynamic systems. As reported by Zhang et al., 2011, this method was initially conceived by Claude E. Shannon in 1948 and is widely used to solve decision-making problems in various fields of life including engineering, management, economics, and medicine, *etc*. Equation 3.1 shows the decision matrix D set up with 'm' alternatives and 'n' criteria:

$$D = [D_{ij}] = \begin{bmatrix} d_{11} & d_{12} & & d_{1n} \\ d_{21} & d_{22} & \cdots & d_{2n} \\ \vdots & \vdots & \cdots & \vdots \\ d_{m1} & d_{m2} & & d_{mn} \end{bmatrix} \tag{3.1}$$

The performance value of the alternatives i (i = 1, 2, ... m) corresponding to each criteria j (j = 1, 2, ... n) is denoted by D_{ij}. The next step is to evaluate a normalized decision matrix N_{ij}, as shown in Equation 3.2. The normalization is performed in order to obtain the non-dimensional values of each criterion necessary to compare them.

$$N_{ij} = \frac{d_{ij}}{\sum_{i=1}^{m} d_{ij}} \tag{3.2}$$

The entropy value of criteria can be obtained as:

$$e_j = -k \sum_{i=1}^{m} N_{ij} \ln(N_{ij}) \quad i = 1, 2, \ldots. m \text{ and } j = 1, 2, \ldots. n \tag{3.3}$$

Where k = 1/ln (m) is a constant that ensures that the value of e_j is between 0 and 1. At this point, the degree of diversification dd_j (Zou et al., 2006) for every criterion is calculated using Equation 3.4:

$$dd_j = |1 - e_j| \tag{3.4}$$

Finally, the weight of each criterion is calculated using Equation 3.5:

$$w_j = \frac{dd_j}{\sum_{j=1}^{n} dd_j} \quad \text{j = 1, 2 .... n} \tag{3.5}$$

3.2.2.2 TOPSIS Method

The TOPSIS is a numerical method that works on the principle of ED (Beltrán & Martínez-Gómez, 2019; Oluah et al., 2020). Hwang & Yoon, 1981 developed the conceptual framework of this method. As per the methodology, the ED will be minimum from a positive (best) ideal solution and maximum from a negative (worst) ideal solution for the best alternative. The solution process starts with the normalization of the decision matrix in Equation 3.1, using Equation 3.6 accordingly:

$$N_{T_{ij}} = \frac{d_{ij}}{\sqrt{\sum_{i=1}^{m} d_{ij}^2}} \quad \text{i = 1, 2, ... m and j = 1, 2, ... n} \tag{3.6}$$

Next, the normalized weighted decision matrix is calculated using the weight of criteria as follows:

$$v_{ij} = w_j \times N_{T_{ij}} \tag{3.7}$$

Next, the ED for each alternative is calculated using Equations 3.8 and 3.9 from the ideal best and worst values of the system under investigation.

$$D_i^* = \sqrt{\sum_{i=1}^{m} (v_{ij} - v*)^2} \quad \text{j = 1, 2 .... n} \tag{3.8}$$

$$D_i^- = \sqrt{\sum_{i=1}^{m} (v_{ij} - v^-)^2} \quad \text{j = 1, 2 .... n} \tag{3.9}$$

At this point, the $CC_{i,}$ i.e. relative closeness index to the ideal value, is calculated, and it lies between 0 and 1. The alternatives are then sorted based on the descending values of the closeness index (i.e. the larger the better):

$$CC_i = \frac{D_i^-}{D_i^* + D_i^-} \tag{3.10}$$

3.2.2.3 VIKOR Method

VIKOR is also an MCDM ranking method that assumes that conflict resolution compromise is bankable. The ultimate aim is to determine the solution which is nearest to the ideal and based on well-established methodology (Beltrán & Martínez-Gómez, 2019; Fernandez et al., 2010; Khare et al., 2013; Wang et al., 2015). All of the alternatives are assessed as per the following procedure: The normalized decision matrix formulated in Equation 3.6 is used to determine the best and the worst values of all criterion functions:

$$N_i^* = \begin{Bmatrix} Max_i N_{ij}\, for\ benefit\ criterion \\ Min_i N_{ij}\, for\ cost\ criterion \end{Bmatrix} \quad i = 1,\ 2 \ldots .\ m \tag{3.11}$$

$$N_i^- = \begin{Bmatrix} Min_i N_{ij}\, for\ benefit\ criterion \\ Max_i N_{ij}\, for\ costcriterion \end{Bmatrix} \quad i = 1,\ 2 \ldots .\ m \tag{3.12}$$

Afterward, regret and utility measures are calculated for all of the alternative materials using Equations 3.13 and 3.14:

$$R_i = Max_i \left[\frac{w_j (N_i^* - N_{ij})}{N_i^* - N_i^-} \right] \tag{3.13}$$

$$S_i = \sum_{j=1}^{n} \left[\frac{w_j (N_i^* - N_{ij}}{N_i^* - N_i^-} \right] \tag{3.14}$$

Finally, the ranking index Q_i for a given v is calculated using Equation 3.15:

$$Q_i = \frac{\nu(S_i - S^*)}{(S^- - S^*)} + \frac{(1 - \nu)(R_i - R^*)}{(R^- - R^*)} \tag{3.15}$$

Where $S^* = min_i S_i$, $S^- = max_i S_i$ and $R^* = min_i R_i$, $R^- = max_i R_i$ denotes the minimum and maximum group utility and regret measures, respectively.

3.3 FINDINGS AND DISCUSSION

3.3.1 Criteria Weighting

The weight of each criterion was obtained, as explained in Section 3.2.2.1, by comparing the properties of each alternative are shown in Annexure A-1. The weights were calculated for six criteria, i.e. melting temperature (T_m), thermal conductivity (k), density (ρ), specific heat capacity (c_p), latent heat (L), and kinematic viscosity (ν). The highest of the weights was assigned to kinematic viscosity (ν), followed by latent heat (L), thermal conductivity (k), melting temperature (T_m), specific heat capacity (c_p), and density (ρ) in decreasing order. The calculated weights are depicted in Figure 3.1. The calculated weights, therefore, were used to implement the TOPSIS and VIKOR methodologies.

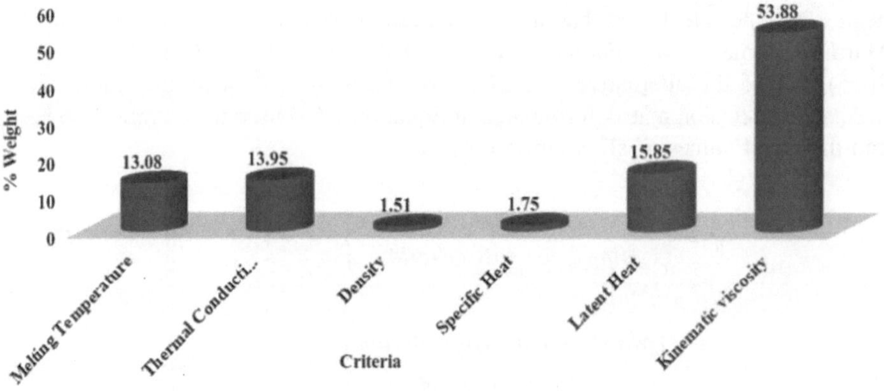

FIGURE 3.1 Weight Distribution for Selected Criteria.

3.3.2 TOPSIS Method

Following the procedure depicted in Section 3.2.2.2, the decision matrix, normalized decision matrix, and weighted matrix were obtained using the criteria weights. The Euclidean distances (ED) were calculated to attain positive and negative ideal solutions. The calculated distances were then used to obtain the relative closeness index (CC_i) and rankings, as shown in Annexure A-2.

The calculated ED are compared and shown in Figure 3.2. This reveals an exciting association among positive ideal (best) and negative (worst) solutions for the PCM alternatives. Moreover, Figure 3.2 also shows the complementary behavior of ED, which indicates the accurate implementation of mathematical operations.

The phase change materials with a performance index near 1.0 represent the positive (best) ideal solution and vice-versa. Fluoranthene, with a performance index of 0.82, was identified as nearest the positive perfect solution, while Ác. Miristico with a performance index of 0.10 was found to be the farthest from the positive ideal solution.

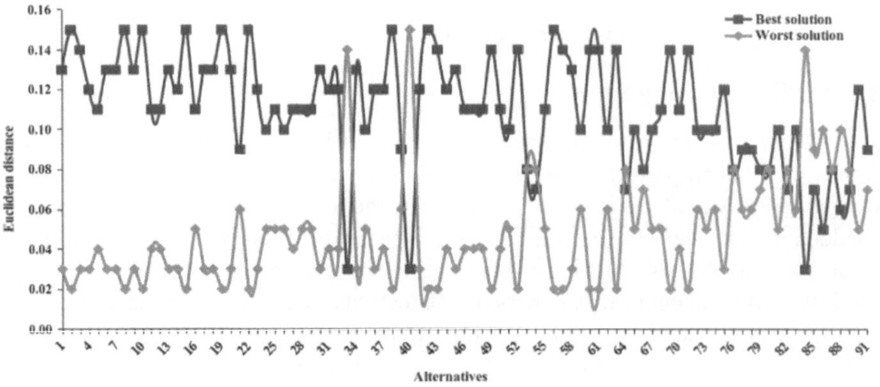

FIGURE 3.2 Comparison Graph of Positive (Best) Ideal and Negative (Worst) ED.

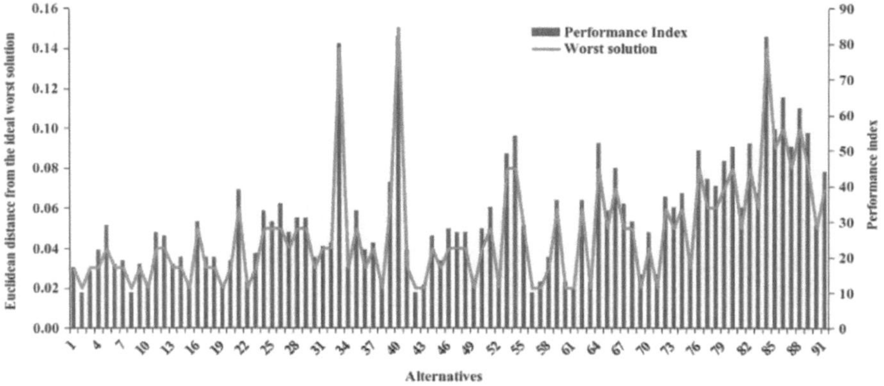

FIGURE 3.3 Graph of Performance Index and ED from the Ideal Worst Solution.

The ED calculated from the negative (worst) ideal solution and the performance index have a direct involvement for the ranking of PCM, as referred to in Equation 3.10. Figure 3.3 shows the plot of the performance index against the ED that are calculated from the ideal worst solution, and this shows an influential association with Equation 3.10. Figure 3.3 indicates that the larger the ED calculated is from the ideal solution, the better and higher the performance index of the PCM are. This denotes that the material which is far away from the ideal negative (worst) solution will be the most suitable material for use. Similar results were presented by Li et al., 2011; Loganathan & Mani, 2018; Oluah et al., 2020; Rastogi et al., 2015; Yang et al., 2018, although their fields of application were different.

3.3.3 VIKOR Method

The values of R_i, S_i, and Q_i were calculated and shown in Annexure A-3. The lowest value of Q_i indicates the best ranking of the material. As per the solution procedure of the VIKOR method presented in Section 3.2.2.3, the ranking of alternatives is illustrated in Annexure A-3.

The calculated regret and utility measures are compared and presented in Figure 3.4; this reveals an exciting association between regret and utility solutions for the PCM alternatives. Figure 3.4 also displays the almost similar behavior of regret and utility measures, which confirms the accurate implementation of mathematical operations.

The PCM with ranking index (Q_i) close to 0 represent a positive ideal solution and vice-versa. Fluoranthene, with a ranking index value of almost 0, was rated the closest to the ideal value and Ác. Láurico, with a ranking index value of 1.0, was ranked the farthest from the ideal value. The ranking index and regret-utility measure have a direct relationship with the ranking of phase change materials, as indicated in Equation 3.15. Figure 3.5 shows the plot of the utility measure against the ranking index and presents an influential association with Equation 3.15. Figure 3.5 expresses that the smaller the regret and utility measures are, the smaller and better the ranking index of the PCM are. It is also evident from Figure 3.5, that

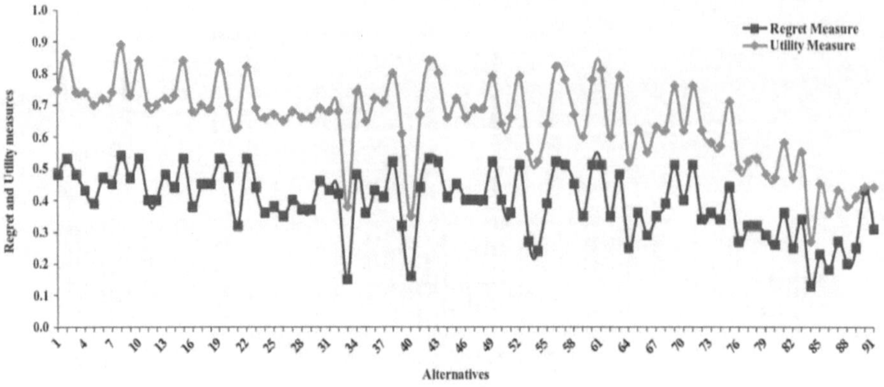

FIGURE 3.4 Comparison of Regret and Utility Measures.

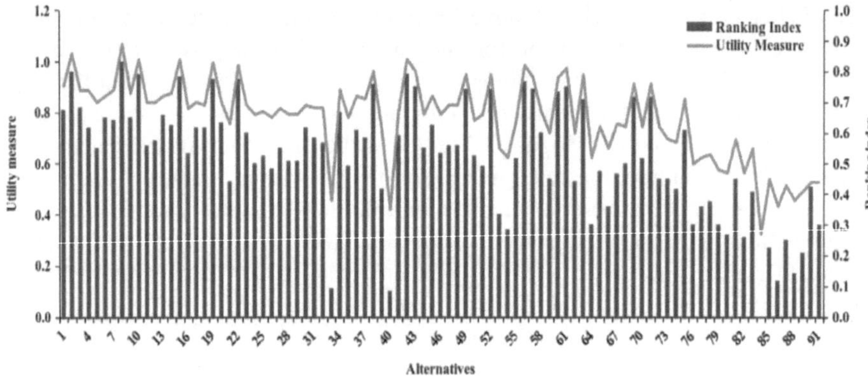

FIGURE 3.5 Graph of Utility Measures and Ranking Index.

the ranking index is a compromised solution between regret and utility measures. This implies that the material with the smallest value of the ranking index will be the most appropriate material for use. Similar observations were presented by Beltrán & Martínez-Gómez, 2019; Wang et al., 2015, although their fields of application were different. Figure 3.6 shows the comparison plot of rankings obtained with the TOPSIS and VIKOR methods. It is clear from Figure 3.6 that both of the processes are in close harmony, with a correlation coefficient of 0.992.

3.4 CONCLUSION AND SUGGESTIONS

Based on the feasibility report presented by the German Technical Cooperation (GTZ), GIZ, and GoI MNRE, 2011 and the literature review, PCM are found to be the promising solution in turning solar energy into a mainstream energy source. For the selection of PCM to use in a solar thermal energy storage system, a selection system was constructed based on the MCDM approach (i.e. Shannon Entropy, TOPSIS, and VIKOR).

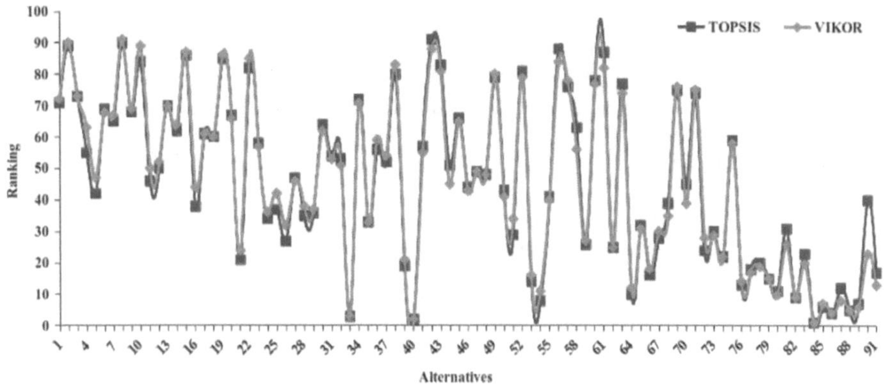

FIGURE 3.6 Comparison Plot of Rankings Obtained with TOPSIS and VIKOR.

Organic PCM available in the literature were evaluated and demonstrated according to the proposed methodology. According to the results of the Shannon Entropy, TOPSIS, and VIKOR methods, the best materials were Fluoranthene, ÁcidotricloroAcético, Succinonitrilo, p-nitrophenol, and flureno. The results show that both TOPSIS and VIKOR could have high acceptance. Based on this preliminary study, the selection of organic PCM for LHTEAS based on the TOPSIS and VIKOR methods is reliable and logical, because the results of both are highly consistent, bearing a correlation coefficient of 0.992. Moreover, selecting the best PCM in order to comply with all of the factors is very difficult. Nonetheless, these two methods, i.e. TOPSIS and VIKOR, are important and necessary for the confident screening and selection of organic PCM.

Moreover, experimental or numerical confirmation is necessary when doing experiments at optimum settings with different weight values assigned to the criteria. The demonstrated approach will be helpful in the selection of PCM for the development of solar thermal energy storage systems. The selected PCM will have a wide variety of applications as a solar thermal energy storage system in various industries like textile (finishing), pulp and paper, pharmaceutical, leather, food processing, dairy, textile (spinning and weaving), electroplating/galvanizing, automobiles and agro malls.

REFERENCES

Athawale, V.M., & Chakraborty, S. (2012). Material selection using multi-criteria decision-making methods: A comparative study. *Proceedings of the Institution of Mechanical Engineers, Part L: Journal of Materials: Design and Applications*, *226*(4), 266–285.

Beltrán, R.D., & Martínez-Gómez, J. (2019). Analysis of phase change materials (PCM) for building wallboards based on the effect of environment. *Journal of Building Engineering*, *24*(July), 100726.

Bilardo, M., Fraisse, G., Pailha, M., & Fabrizio, E. (2020). Design and experimental analysis of an integral collector storage (ICS) prototype for DHW production. *Applied Energy*, *259*(July), 114104.

Chakraborty, S., & Chatterjee, P. (2013). Selection of materials using multi-criteria decision-making methods with minimum data. *Decision Science Letters*, (July), *12*(3), 135–148.

Çolak, M., & Kaya, İ. (2020). Multi-criteria evaluation of energy storage technologies based on hesitant fuzzy information: A case study for Turkey. *Journal of Energy Storage*, *28*(July 2019), 101211.

Fernandez, A.I., Martínez, M., Segarra, M., Martorell, I., & Cabeza, L.F. (2010). Selection of materials with potential in sensible thermal energy storage. *Solar Energy Materials and Solar Cells*, *94*(10), 1723–1729.

German Technical Cooperation (GTZ), GIZ, and GoI MNRE. (2011). *Identification of industrial sectors promising for commercialisation of solar energy*. German Technical Cooperation (GTZ) and MNRE (GoI), India.

Hafezalkotob, A., & Hafezalkotob, A. (2016). Extended MULTIMOORA method based on Shannon entropy weight for materials selection. *Journal of Industrial Engineering International*, *12*(1), 1–13.

Hwang, C.-L., & Yoon, K. (1981). Methods for multiple attribute decision making. In Hwang, C.-L. and Yoon, K., *Multiple Attribute Decision Making*. (Lecture Notes in Economics and Mathematical Systems, Vol. 186, pp. 58–191). Springer, Berlin, Heidelberg.

Khare, S., Dell'Amico, M., Knight, C., & McGarry, S. (2013). Selection of materials for high temperature sensible energy storage. *Solar Energy Materials and Solar Cells*, *115*(August), 114–122.

Li, X., Wang, K., Liu, L., Xin, J., Yang, H., & Gao, C. (2011). Application of the entropy weight and TOPSIS method in safety evaluation of coal mines. *Procedia Engineering*, *26*, 2085–2091.

Liu, H.-C., You, J.-X., Zhen, L., & Fan, X.-J. (2014). A novel hybrid multiple criteria decision making model for material selection with target-based criteria. *Materials & Design*, *60*(August), 380–390.

Loganathan, A., & Mani, I. (2018). A fuzzy based hybrid multi criteria decision making methodology for phase change material selection in electronics cooling system. *Ain Shams Engineering Journal*, *9*(4), 2943–2950.

Milani, A. S., Eskicioglu, C., Robles, K., Bujun, K., & Hosseini-Nasab, H. (2011). Multiple criteria decision making with life cycle assessment for material selection of composites. *Express Polymer Letters*, *5*(12), 1062–1074.

Mittal, V. K., Sindhwani, R., Shekhar, H., & Singh, P. L. (2019). Fuzzy AHP model for challenges to thermal power plant establishment in India. *International Journal of Operational Research*, *34*(4), 562–581.

Oluah, C., Akinlabi, E.T., & Njoku, H.O. (2020). Selection of phase change material for improved performance of Trombe wall systems using the entropy weight and TOPSIS methodology. *Energy and Buildings*, *217*(June), 109967.

Pereira da Cunha, J., & Eames, P. (2016). Thermal energy storage for low and medium temperature applications using phase change materials – A review. *Applied Energy*, *177*(September), 227–238.

Rastogi, M., Chauhan, A., Vaish, R., & Kishan, A. (2015). Selection and performance assessment of phase change materials for heating, ventilation and air-conditioning applications. *Energy Conversion and Management*, *89*(January), 260–269.

Rathod, M.K., & Kanzaria, H.V. (2011). A methodological concept for phase change material selection based on multiple criteria decision analysis with and without fuzzy environment. *Materials & Design*, *32*(6), 3578–3585.

Saaty, T.L. (1990). How to make a decision: The analytic hierarchy process. *European Journal of Operational Research*, *48*(1), 9–26.

Sardari, P.T., Grant, D., Giddings, D., Walker, G.S., & Gillott, M. (2019). Composite metal foam/PCM energy store design for dwelling space air heating. *Energy Conversion and Management*, *201*(July), 112–151.

Shanker, K., Shankar, R., & Sindhwani, R. (Eds.). (2019). *Advances in industrial and production engineering: Select proceedings of FLAME 2018*. Springer.

Sindhwani, R., & Malhotra, V. (2017). A framework to enhance agile manufacturing system: A total interpretive structural modelling (TISM) approach. *Benchmarking: An International Journal, 24*(2), 467–487. https://doi.org/10.1108/BIJ-09-2015-0092

Sindhwani, R., Mittal, V.K., Singh, P.L., Kalsariya, V., & Salroo, F. (2018). Modelling and analysis of energy efficiency drivers by fuzzy ISM and fuzzy MICMAC approach. *International Journal of Productivity and Quality Management, 25*(2), 225–244.

Sindhwani, R., Singh, P.L., Chopra, R., Sharma, K., Basu, A., Prajapati, D.K., & Malhotra, V. (2019). Agility evaluation in the rolling industry: A case study. In K. Shanker, R. Shankar, & R. Sindhwani (Eds.), *Advances in Industrial and Production Engineering*. Lecture Notes in Mechanical Engineering (pp. 753–770). Springer, Singapore.

Sindhwani, R., Singh, P.L., Iqbal, A., Prajapati, D.K., & Mittal, V.K. (2019). Modeling and analysis of factors influencing agility in healthcare organizations: An ISM approach. In K. Shanker, R. Shanker, & R. Sindhwani (Eds.), *Advances in Industrial and Production Engineering*. Lecture Notes in Mechanical Engineering (pp. 683–696). Springer, Singapore. https://doi.org/10.1007/978-981-13-6412-9_64

Singh, R. P., Xu, H., Kaushik, S.C., Rakshit, D., & Romagnoli, A. (2019). Charging performance evaluation of finned conical thermal storage system encapsulated with nano-enhanced phase change material. *Applied Thermal Engineering, 151*(September 2018), 176–190.

Venkata Rao, R. (2009). An improved compromise ranking method for evaluation of environmentally conscious manufacturing programs. *International Journal of Production Research, 47*(16), 4399–4412.

Vitorino, N., Abrantes, J.C.C., & Frade, J.R. (2016). Quality criteria for phase change materials selection. *Energy Conversion and Management, 124*(September), 598–606.

Wang, Y., Zhang, Y., Yang, W., & Ji, H. (2015). Selection of low-temperature phase-change materials for thermal energy storage based on the VIKOR method. *Energy Technology, 3*(1), 84–89.

Wei, G., Wang, G., Xu, C., Ju, X., Xing, L., Du, X., & Yang, Y. (2018). Selection principles and thermophysical properties of high temperature phase change materials for thermal energy storage: A review. *Renewable and Sustainable Energy Reviews, 81*(January), 1771–1786.

Weng, J., He, Y., Ouyang, D., Yang, X., Zhang, G., & Wang, J. (2019). Thermal performance of PCM and branch-structured fins for cylindrical power battery in a high-temperature environment. *Energy Conversion and Management, 200*(September), 112106.

Xu, H., Sze, J.Y., Romagnoli, A., & Py, X. (2017). Selection of phase change material for thermal energy storage in solar air conditioning systems. *Energy Procedia, 105*, 4281–4288.

Yang, K., Zhu, N., Chang, C., Wang, D., Yang, S., & Ma, S. (2018). A methodological concept for phase change material selection based on multi-criteria decision making (MCDM): A case study. *Energy, 165*(December), 1085–1096.

Yazdani, M., & FarokhPayam, A. (2015). A comparative study on material selection of microelectromechanical systems electrostatic actuators using Ashby, VIKOR and TOPSIS. *Materials & Design (1980–2015), 65*(January), 328–334.

Zayed, M.E., Zhao, J., Elsheikh, A.H., Hammad, F.A., Ma, L., Du, Y., … Shalaby, S.M. (2019). Applications of cascaded phase change materials in solar water collector storage tanks: A review. *Solar Energy Materials and Solar Cells, 199*(December 2018), 24–49.

Zhang, H., Gu, C.-L., Gu, L.-W., & Zhang, Y. (2011). The evaluation of tourism destination competitiveness by TOPSIS & information entropy – A case in the Yangtze River Delta of China'. *Tourism Management, 32*(2), 443–451.

Zou, Z.-H., Yun, Y., & Sun, J.-N. (2006). Entropy method for determination of weight of evaluating indicators in fuzzy synthetic evaluation for water quality assessment. *Journal of Environmental Sciences, 18*(5), 1020–1023.

ANNEXURES A-1: THERMO-PHYSICAL PROPERTIES OF THE CANDIDATE ORGANIC PCM

S. No.	Material	T_m (°C)	k (W/mK)	ρ (kg/m^3)	Cp (kJ/kg°C)	L (kJ/kg)	ν (m^2/s)
1	Docasyl bromide	40	0.5	900	1.8	201	1.10E-07
2	Parafina C21	40.2	0.2	900	2	200	4.70E-08
3	1-ciclohexilooc-tadecano	41	0.5	900	1.8	218	1.00E-07
4	Fenol	41	0.5	1,000	1.8	120	1.70E-07
5	P-clorophenol	42.7	0.5	900	2	109.5	2.20E-07
6	Ácidododecanoico	43.2	0.5	865	2.3	208.2	1.20E-07
7	P-metilanilina	43.8	0.5	900	1.2	170.3	1.40E-07
8	Ác. Láurico	44	0.1	1,007	1.7	211.6	3.00E-08
9	Cianamida	44	0.5	900	1.8	209	1.20E-07
10	Parafina C22	44	0.2	900	2	249	4.10E-08
11	M-cloronitro-benzeno	44.4	0.5	850	2	123	2.10E-07
12	O-nitrophenol	44.8	0.5	900	2	127.3	2.00E-07
13	Methyl eicosate	45	0.5	900	1.8	230	1.10E-07
14	2,6-Xilenol	45.7	0.5	965	2	154.9	1.50E-07
15	Parafina C23	47.5	0.2	930	2	232	4.60E-08
16	Benzophenol	47.9	0.5	900	2	115.1	2.30E-07
17	2-heptadecanona	48	0.5	805	2	218	1.40E-07
18	3-Heptadecanona	48	0.5	805	2.5	218	1.40E-07
19	Parafina C25	49.4	0.2	930	2.2	238	4.70E-08
20	Camfeno	50	0.5	900	1.8	238	1.20E-07
21	O-nitroanilina	50	0.5	900	2.3	93	3.00E-07
22	Parafina C24	50.6	0.2	930	2	255	4.50E-08
23	9-heptadecanona	51	0.5	814	2	213	1.50E-07
24	Timol	51.5	0.5	900	2.3	115	2.50E-07
25	P-Nitrotolueno	51.6	0.5	900	2	122.7	2.30E-07
26	P-diclorobenzeno	52.7	0.5	850	1.1	116.7	2.70E-07
27	Difenilamina	52.9	0.5	1,160	2.3	107	2.10E-07
28	Diphenilamina	53	0.5	900	2	123.2	2.40E-07
29	P-diclorobenzeno	53.1	0.5	900	2.3	121	2.40E-07
30	Ácidotetrade-canóico	54	0.5	900	2.3	224.7	1.30E-07
31	Oxolato	54.3	0.5	900	2.3	178	1.70E-07
32	Dimetiloxalato	54.4	0.5	850	2	178.6	1.80E-07
33	Succinonitrilo	54.5	0.5	850	1.8	49	6.50E-07
34	Ác. Hipofosfórico	55	0.4	900	2	213	1.10E-07
35	O-xilenodic-loridrico	55	0.5	900	2.3	121	2.50E-07

(*Continued*)

S. No.	Material	T_m (°C)	k (W/mK)	ρ (kg/m^3)	Cp (kJ/kg°C)	L (kJ/kg)	ν (m^2/s)
36	Ác. Beta-cloroacético	56	0.4	900	2	147	1.70E-07
37	Ác. Cloroacético	56	0.4	900	2	130	1.90E-07
38	Parafina C26	56.3	0.2	900	2.2	256	5.10E-08
39	Nitro naftaleno	56.7	0.5	900	2.3	103	3.10E-07
40	Ácidotricloro-Acético	57.5	0.4	900	2	36	7.10E-07
41	1-Hexadecanol	58	0.5	900	2	220	1.50E-07
42	Ác. Miristico	58	0.2	990	1.7	186.6	4.70E-08
43	Parafina C27	58.8	0.2	880	2.2	236	5.90E-08
44	1,3,5-Trioxano	60.2	0.5	900	2	179.9	1.90E-07
45	Ác. Heptadecanóico	60.6	0.4	900	2	189	1.40E-07
46	3,4-Xilenol	60.8	0.5	983	2	148.6	2.10E-07
47	Ác. Alpha cloroacético	61.2	0.4	900	2	130	2.10E-07
48	ÁcicoCloroAcético	61.3	0.4	900	2	129.9	2.10E-07
49	Parafina C28	61.6	0.2	880	2.2	253	5.80E-08
50	Ácidohexade-canóico	61.8	0.5	900	2.3	164.2	2.10E-07
51	Ác. Glicólico	63	0.4	900	2	109	2.60E-07
52	Parafina C29	63.4	0.2	880	2	240	6.30E-08
53	1,3,5-Tricloro-benzeno	63.5	0.5	850	2	100.3	3.70E-07
54	P-bromofenol	63.5	0.5	900	2.3	86	4.10E-07
55	3,5-Xilenol	63.6	0.5	968	2	147.5	2.20E-07
56	Ác. Palmítico	64	0.2	989	1.9	185.4	5.70E-08
57	Parafina C30	65.4	0.2	900	2	251	6.10E-08
58	Poliglicol E6000	66	0.5	1,212	2.3	190	1.40E-07
59	AzoBenzeno	67.1	0.5	900	2	139.5	2.70E-07
60	Parafina C31	68	0.2	900	2	242	6.60E-08
61	Ác. Esteárico	69	0.2	965	1.6	202.5	6.10E-08
62	Biphenil	69	0.5	900	2	143.1	2.70E-07
63	Parafina C32	69.5	0.2	850	1.8	170	1.00E-07
64	Dintotolueno (2,4)	70	0.5	900	2.3	101	3.90E-07
65	Biphenyl	71	0.5	1,166	2	119.2	2.50E-07
66	O-nitroanilina	71.2	0.5	900	2	116.7	3.40E-07
67	Ácido trans-crotónico	72	0.4	990	2.1	106	2.70E-07
68	2,3-Xilenol	72.8	0.5	965	2	172.3	2.20E-07
69	Parafina C33	73.9	0.2	900	1.8	268	6.40E-08
70	2,5-Xilenol	74.8	0.5	965	2	191.6	2.00E-07

(*Continued*)

S. No.	Material	T_m (°C)	k (W/mK)	ρ (kg/m^3)	Cp (kJ/kg°C)	L (kJ/kg)	ν (m^2/s)
71	Parafina C34	75.9	0.2	900	1.8	269	6.60E-08
72	Ác. Fenilacético	76.7	0.4	1,081	2	102	2.80E-07
73	Benzilamina	78	0.5	900	2	174	2.50E-07
74	1,2,4,5-Tetrametil-benzeno	79.3	0.5	900	1.8	156.7	2.80E-07
75	Naftaleno	80	0.3	1,145	1.9	148	1.60E-07
76	P-cloronitrobenzeno	83.5	0.5	850	2	131.9	3.70E-07
77	Acnaphtaleno	93.4	0.5	900	2	165.7	3.10E-07
78	Benzil	94.9	0.5	1,230	2	126.6	3.00E-07
79	1-Naphtol	96	0.5	850	2	162	3.50E-07
80	M-nitrophenol	96.8	0.5	900	2	140.1	3.80E-07
81	ÁcidoGlutaric	97.8	0.4	990	2.1	158.3	2.50E-07
82	Fenantreno	99.2	0.5	1,179	2	106.9	3.90E-07
83	Ácido o-Toluic	104	0.4	990	2.1	148.3	2.80E-07
84	Fluoranthene	108	0.5	900	2	93.4	6.40E-07
85	ácido m-Toluic	109	0.4	900	2.1	115.6	4.20E-07
86	P-nitrophenol	114	0.5	900	2	133.2	4.80E-07
87	M-nitroanilina	114	0.5	900	2	171.6	3.70E-07
88	Flureno	115	0.5	900	2	137.9	4.60E-07
89	P-benzoquinona	116	0.5	850	2	171.6	4.00E-07
90	Erythriol	118	0.7	1,480	1.4	339.8	1.70E-07
91	Succinic anhydride	119	0.5	850	2	217.1	3.20E-07

ANNEXURE A-2: RANKING OF CANDIDATE ORGANIC PCM BASED ON THE TOPSIS METHOD

S. No.	Material	Positive Ideal Solution	Negative Ideal Solution	Closeness Index	Ranking
1	Docasyl bromide	0.13	0.03	0.17	71
2	Parafina C21	0.15	0.02	0.10	89
3	1-ciclohexilooctadecano	0.14	0.03	0.16	73
4	Fenol	0.12	0.03	0.22	55
5	P-clorophenol	0.11	0.04	0.29	42
6	Ácidododecanoico	0.13	0.03	0.18	69
7	P-metilanilina	0.13	0.03	0.19	65
8	Ác. Láurico	0.15	0.02	0.10	90
9	Cianamida	0.13	0.03	0.18	68
10	Parafina C22	0.15	0.02	0.12	84
11	M-cloronitrobenzeno	0.11	0.04	0.27	46
12	O-nitrophenol	0.11	0.04	0.26	50

(*Continued*)

S. No.	Material	Positive Ideal Solution	Negative Ideal Solution	Closeness Index	Ranking
13	Methyl eicosate	0.13	0.03	0.18	70
14	2,6-Xilenol	0.12	0.03	0.20	62
15	Parafina C23	0.15	0.02	0.12	86
16	Benzophenol	0.11	0.05	0.30	38
17	2-heptadecanona	0.13	0.03	0.20	61
18	3-Heptadecanona	0.13	0.03	0.20	60
19	Parafina C25	0.15	0.02	0.12	85
20	Camfeno	0.13	0.03	0.19	67
21	O-nitroanilina	0.09	0.06	0.39	21
22	Parafina C24	0.15	0.02	0.13	82
23	9-heptadecanona	0.12	0.03	0.21	58
24	Timol	0.10	0.05	0.33	34
25	P-Nitrotolueno	0.11	0.05	0.30	37
26	P-diclorobenzeno	0.10	0.05	0.35	27
27	Difenilamina	0.11	0.04	0.27	47
28	Diphenilamina	0.11	0.05	0.31	35
29	P-diclorobenzeno	0.11	0.05	0.31	36
30	Ácidotetradecanóico	0.13	0.03	0.20	64
31	Oxolato	0.12	0.04	0.23	54
32	Dimetiloxalato	0.12	0.04	0.24	53
33	Succinonitrilo	0.03	0.14	0.80	3
34	Ác. Hipofosfórico	0.13	0.03	0.17	72
35	O-xilenodicloridrico	0.10	0.05	0.33	33
36	Ác. Beta-cloroacético	0.12	0.03	0.22	56
37	Ác. Cloroacético	0.12	0.04	0.24	52
38	Parafina C26	0.15	0.02	0.13	80
39	Nitro naftaleno	0.09	0.06	0.41	19
40	ÁcidotricloroAcético	0.03	0.15	0.82	2
41	1-Hexadecanol	0.12	0.03	0.22	57
42	Ác. Miristico	0.15	0.02	0.10	91
43	Parafina C27	0.14	0.02	0.12	83
44	1,3,5-Trioxano	0.12	0.04	0.26	51
45	Ác. Heptadecanóico	0.13	0.03	0.19	66
46	3,4-Xilenol	0.11	0.04	0.28	44
47	Ác. Alpha cloroacético	0.11	0.04	0.27	49
48	ÁcicoCloroAcético	0.11	0.04	0.27	48
49	Parafina C28	0.14	0.02	0.13	79
50	Ácidohexadecanóico	0.11	0.04	0.28	43
51	Ác. Glicólico	0.10	0.05	0.34	29
52	Parafina C29	0.14	0.02	0.13	81
53	1,3,5-Triclorobenzeno	0.08	0.08	0.49	14
54	P-bromofenol	0.07	0.08	0.54	8

(Continued)

S. No.	Material	Positive Ideal Solution	Negative Ideal Solution	Closeness Index	Ranking
55	3,5-Xilenol	0.11	0.05	0.29	41
56	Ác. Palmítico	0.15	0.02	0.10	88
57	Parafina C30	0.14	0.02	0.13	76
58	Poliglicol E6000	0.13	0.03	0.20	63
59	AzoBenzeno	0.10	0.06	0.36	26
60	Parafina C31	0.14	0.02	0.13	78
61	Ác. Esteárico	0.14	0.02	0.11	87
62	Biphenil	0.10	0.06	0.36	25
63	Parafina C32	0.14	0.02	0.13	77
64	Dintotolueno (2,4)	0.07	0.08	0.52	10
65	Biphenyl	0.10	0.05	0.33	32
66	O-nitroanilina	0.08	0.07	0.45	16
67	ácido trans-crotónico	0.10	0.05	0.35	28
68	2,3-Xilenol	0.11	0.05	0.30	39
69	Parafina C33	0.14	0.02	0.15	75
70	2,5-Xilenol	0.11	0.04	0.27	45
71	Parafina C34	0.14	0.02	0.15	74
72	Ác. Fenilacético	0.10	0.06	0.37	24
73	Benzilamina	0.10	0.05	0.34	30
74	1,2,4,5-Tetrametilbenzeno	0.10	0.06	0.38	22
75	Naftaleno	0.12	0.03	0.21	59
76	P-cloronitrobenzeno	0.08	0.08	0.50	13
77	Acnaphtaleno	0.09	0.06	0.42	18
78	Benzil	0.09	0.06	0.40	20
79	1-Naphtol	0.08	0.07	0.47	15
80	M-nitrophenol	0.08	0.08	0.51	11
81	ÁcidoGlutaric	0.10	0.05	0.34	31
82	Fenantreno	0.07	0.08	0.52	9
83	Ácido o-Toluic	0.10	0.06	0.38	23
84	Fluoranthene	0.03	0.14	0.82	1
85	Ácido m-Toluic	0.07	0.09	0.56	6
86	P-nitrophenol	0.05	0.10	0.65	4
87	M-nitroanilina	0.08	0.08	0.51	12
88	Flureno	0.06	0.10	0.62	5
89	P-benzoquinona	0.07	0.08	0.55	7
90	Erythriol	0.12	0.05	0.29	40
91	Succinic anhydride	0.09	0.07	0.44	17

ANNEXURE A-3: RANKING OF CANDIDATE ORGANIC PCM BASED ON THE VIKOR METHOD

S. No.	Material	Regret Measure	Utility Measure	Ranking Index	Ranking
1	Docasyl bromide	0.48	0.75	0.81	72
2	Parafina C21	0.53	0.86	0.96	90
3	1-ciclohexilooctadecano	0.48	0.74	0.82	73
4	Fenol	0.43	0.74	0.74	63
5	P-clorophenol	0.39	0.7	0.66	47
6	Ácidododecanoico	0.47	0.72	0.78	68
7	P-metilanilina	0.45	0.74	0.77	67
8	Ác. Láurico	0.54	0.89	1	91
9	Cianamida	0.47	0.73	0.78	69
10	Parafina C22	0.53	0.84	0.95	89
11	M-cloronitrobenzeno	0.4	0.7	0.67	50
12	O-nitrophenol	0.4	0.7	0.69	52
13	Methyl eicosate	0.48	0.72	0.79	70
14	2,6-Xilenol	0.44	0.73	0.75	64
15	Parafina C23	0.53	0.84	0.94	87
16	Benzophenol	0.38	0.68	0.64	44
17	2-heptadecanona	0.45	0.7	0.74	61
18	3-Heptadecanona	0.45	0.69	0.74	60
19	Parafina C25	0.53	0.83	0.93	86
20	Camfeno	0.47	0.7	0.76	66
21	O-nitroanilina	0.32	0.63	0.53	24
22	Parafina C24	0.53	0.82	0.93	85
23	9-heptadecanona	0.44	0.69	0.72	57
24	Timol	0.36	0.66	0.6	36
25	P-Nitrotolueno	0.38	0.67	0.63	42
26	P-diclorobenzeno	0.35	0.65	0.58	32
27	Difenilamina	0.4	0.68	0.66	46
28	Diphenilamina	0.37	0.66	0.61	38
29	P-diclorobenzeno	0.37	0.66	0.61	37
30	Ácidotetradecanóico	0.46	0.69	0.74	62
31	Oxolato	0.43	0.68	0.7	53
32	Dimetiloxalato	0.42	0.68	0.68	51
33	Succinonitrilo	0.15	0.38	0.11	3
34	Ác. Hipofosfórico	0.48	0.74	0.8	71
35	O-xilenodicloridrico	0.36	0.65	0.59	33
36	Ác. Beta-cloroacético	0.43	0.72	0.73	59
37	Ác. Cloroacético	0.41	0.71	0.7	54
38	Parafina C26	0.52	0.8	0.91	83

(*Continued*)

S. No.	Material	Regret Measure	Utility Measure	Ranking Index	Ranking
39	Nitro naftaleno	0.32	0.61	0.5	21
40	ÁcidotricloroAcético	0.16	0.35	0.1	2
41	1-Hexadecanol	0.44	0.67	0.71	55
42	Ác. Miristico	0.53	0.84	0.95	88
43	Parafina C27	0.52	0.8	0.9	81
44	1,3,5-Trioxano	0.41	0.66	0.66	45
45	Ác. Heptadecanóico	0.45	0.72	0.75	65
46	3,4-Xilenol	0.4	0.66	0.64	43
47	Ác. Alpha cloroacético	0.4	0.69	0.67	49
48	ÁcicoCloroAcético	0.4	0.69	0.67	48
49	Parafina C28	0.52	0.79	0.89	80
50	Ácidohexadecanóico	0.4	0.64	0.63	41
51	Ác. Glicólico	0.36	0.66	0.59	34
52	Parafina C29	0.51	0.79	0.89	79
53	1,3,5-Triclorobenzeno	0.27	0.55	0.4	16
54	P-bromofenol	0.24	0.52	0.34	11
55	3,5-Xilenol	0.39	0.64	0.62	40
56	Ác. Palmítico	0.52	0.82	0.92	84
57	Parafina C30	0.51	0.78	0.89	78
58	Poliglicol E6000	0.45	0.67	0.72	56
59	AzoBenzeno	0.35	0.6	0.54	27
60	Parafina C31	0.51	0.78	0.88	77
61	Ác. Esteárico	0.51	0.81	0.9	82
62	Biphenil	0.35	0.6	0.53	25
63	Parafina C32	0.48	0.79	0.85	74
64	Dintotolueno (2,4)	0.25	0.52	0.36	12
65	Biphenyl	0.36	0.62	0.57	31
66	o-nitroanilina	0.29	0.55	0.43	18
67	Ácido trans-crotónico	0.35	0.63	0.56	30
68	2,3-Xilenol	0.39	0.62	0.6	35
69	Parafina C33	0.51	0.76	0.86	76
70	2,5-Xilenol	0.4	0.62	0.62	39
71	Parafina C34	0.51	0.76	0.86	75
72	Ác. Fenilacético	0.34	0.62	0.54	28
73	Benzilamina	0.36	0.58	0.54	29
74	1,2,4,5-Tetrametilbenzeno	0.34	0.57	0.5	22
75	Naftaleno	0.44	0.71	0.73	58
76	P-cloronitrobenzeno	0.27	0.5	0.36	14
77	Acnaphtaleno	0.32	0.52	0.43	17
78	Benzil	0.32	0.53	0.45	19
79	1-Naphtol	0.29	0.48	0.36	15

(Continued)

S. No.	Material	Regret Measure	Utility Measure	Ranking Index	Ranking
80	M-nitrophenol	0.26	0.47	0.32	10
81	ÁcidoGlutaric	0.36	0.58	0.54	26
82	Fenantreno	0.25	0.47	0.31	9
83	Ácido o-Toluic	0.34	0.55	0.49	20
84	Fluoranthene	0.13	0.27	0	1
85	Ácido m-Toluic	0.23	0.45	0.27	7
86	P-nitrophenol	0.18	0.36	0.14	4
87	M-nitroanilina	0.27	0.43	0.3	8
88	Flureno	0.2	0.38	0.17	5
89	P-benzoquinona	0.25	0.41	0.25	6
90	Erythriol	0.43	0.44	0.51	23
91	Succinic anhydride	0.31	0.44	0.36	13

4 Identification and Analysis of Enablers of Socially Sustainable SMEs Using the MCDM Approach

Shubhanshi Mittal, Shreya Gupta, and Vernika Agarwal

CONTENTS

4.1 Introduction 65
4.2 Review of Literature 67
4.2.1 SMEs 67
4.2.2 ISM Application 68
4.2.3 Economic, Innovation, and Leadership Dimensions 69
4.3 Research Methodology 69
4.3.1 Identification of Factors 69
4.3.2 ISM Methodology 71
4.4 Case Background 73
4.4.1 Case Description 74
4.5 Results and Discussion 75
4.6 Conclusion 78
4.7 Managerial Implication 79
References 79

4.1 INTRODUCTION

India's Small and Medium Enterprises (SMEs) perform a vital part in enhancing and strengthening the overall financial backbone of the country (Centobelli et al., 2019). After globalization, SMEs are needed to pursue international and domestic markets. Globalization has forced SMEs to provide products with good quality, reduced cost, and a variety of products with higher performance as well as improved technology (Singh et al., 2007). With globalization, technological changes have enhanced performances and physical resources to generate ideas that have helped in translating these to social and economic value (Kumar et al., 2017c; Narula, 2004). SMEs played

TABLE 4.1
Prominent Government Schemes for MSMEs (Source: msme.gov.in)

Name of the Scheme	Year	Description
Prime Minister Employment Generation Program (PMEGP)	2008	Employment opportunity generation through aid in setting up self-employment ventures.
Market Promotion and Development Scheme (MPDA)	2010	Marketing and promotion of khadi and village SMEs.
Lean Manufacturing Competitiveness for MSMEs	2015	Manufacturing competitiveness upgrade of MSMEs through lean manufacturing techniques.
Skill Upgradation and Mahila Coir Yojana (MCY)	2019–2020	Growth of artisan women promotion via self-employment opportunities.

a significant part in goals like equality of income, poverty, regional imbalances, *etc.* With the help of SMEs, there are over 6,000 products being manufactured, including items from daily consumable goods to high-tech equipment. Also, the results show that SMEs are proving efficacious in distributing employment opportunities at a minimal cost compared to large-scale enterprises. The Confederation of Indian Industry (CII) said that in the manufacturing sector SMEs contribute about 6.11% GDP and 24.6% GDP (service sector) and generate about 80 million jobs; hence. they play a crucial role in India's economy. Accordingly, the focus of the authority of the country in transferring towards the development of SMEs has improved. The initiatives are supplementary in the consolidation of SME organisations. Some schemes offered by the government of India are listed in Table 4.1.

Besides these schemes, the SME Chamber of India has laid down an action plan which majorly includes:

- Transformation as well as the transition of SMEs.
- Focus on the ease of doing business for SMEs.
- Reduced rate of interest for MSME sectors.
- Industrial plots in prime locations allocated to SMEs.
- Special power tariff for SMEs.

Mani et al. (2020) proposed that there is an affiliation between social sustainability in SMEs and the performance supply chain. As large-scale businesses are going towards sustainable growth to progress their efficiency and save energy, stakeholders are likewise pushing the SMEs towards sustainability. Sustainability in SMEs helps in preserving the changing world and balancing business from economic, social, legal, and ecological impact (Kumar et al., 2017b; Mittal et al., 2019). SMEs can adopt sustainable development easily as contrasted to big firms, as SMEs have a much simpler structure. Sustainability programs help SMEs to save in terms of reduction of energy and water consumption, which further help to reduce cost and increase profits (Meath et al., 2016; Sindhwani et al., 2018).

Research has scrutinized that there are additional influences to SMEs, apart from the government intervention, that has an impact on the SME green expansion, such as Entrepreneurial Orientation (EO) in times of recession which have contradictory effects on the operations of SMEs (Soininen et al., 2012). Information and Communication Technologies (ICT) impact the financial, societal, and personal development of SMEs, and it is, hence, vital to use potential resources the most (Tarutė & Gatautis, 2014). Political economy is a major way of evaluation in comprehending the bearing of SMEs' policies. The current targets and fallouts of enterprises are achieved with the help of policy effectiveness (Kumar, 2017a; Storey, 2006). Entrepreneurial knowledge and presentation have a progressive effect on SMEs and their dimensions. (Omerzel & Antončič, 2008). Moreover, research done on transnational corporations can help in high-tech improvement, which will then result in improved cost-effective performance (Kumar & Subrahmanya, 2010). It is evident from the existing studies by numerous researchers in the literature that the central focus has been on the economic aspect for the functioning of SMEs, and not much has been done in enhancing the sustainability of SMEs. In this study, the target is to integrate social sustainability with the legal and business dimensions of SMEs in the northern part of India.

Thus, the objectives of the current study are as follows:

- To recognize the enablers for acceptance of social sustainability through a thorough literature survey.
- To gauge the interrelationships between the different enablers of social sustainability.
- To improve a functioning model for the enablers of social sustainability using ISM in order to understand communications among several pressures.
- To validate the developed model with the help of domain experts.
- To classify the different groups of enablers on the basis of their driving power and dependence using the MICMAC analysis.

Socially, sustainability has now become a principal focus of many enterprises, as human rights are the first of six United Nations Global Compact's social dimensions of corporate sustainability. Furthermore, indulging in social sustainability can open new niches for enterprises or businesses, like attracting new potential investors or clients, which is the source for innovation of new products or services.

In order to tackle social sustainability encompassing the legal and business dimensions of SMEs, enablers were identified with the help of industry professionals and scholars. The study is done using the Multi-Criteria Decision-Making approach of Interpretive Structural Modelling (ISM).

4.2 REVIEW OF LITERATURE

4.2.1 SMEs

The case of SMEs differs from that of large firms, as they require particular attention to green operations because of the increasing pressure from stakeholders (Stubblefield Loucks et al., 2010). SMEs require different combinations of resources like equity, research and development, industry knowledge, and expertise to move towards sustainable development (Halme & Korpela, 2014). To goal to be sustainable and the management of social sustainability are challenges for these SMEs, as they must gain knowledge and expertise to properly implement and manage their human resources while abiding with government norms. One way to achieve social sustainability is by including internal cooperation of the top body assurance and external infrastructure (Johnson, 2017). The top supervision is mainly accountable for corporate social responsibility in an organisation. It was established that rivalry and corporate social responsibility are co-related and proven as barriers (Battaglia et al., 2014).

The key challenges faced by SMEs in their goal for achieving a sustainable enterprise include lack of resources, high capital and operating costs, and inadequacy of expertise (Álvarez Jaramillo et al., 2018) Broadly, 175 barriers faced by SMEs were identified via research equations and results in the Scopus of previous studies (Álvarez Jaramillo et al., 2018). In India, manufacturing SMEs contribute a significant amount of shares when compared with the manufacturing output of the country. Enterprises should use lean and green manufacturing techniques to achieve desired results (Gandhi et al., 2018). In this context, the prime step for Indian manufacturing SMEs is to reduce wastage by opting for lean manufacturing (Panizzolo et al., 2012, Rizos et al., 2016). SMEs highly contribute to the worldwide economy. In the study, sustainable supply chain management (SCM) has a rather perilous role in terms of business, environmental, and social aspects (Kot, 2018). Government intervention in terms of policy and environmental practices positively influences the performances of Indian leather and chemical SMEs (Das, 2019). In the literature, it can be perceived that there is relatively limited work done in the field of social sustainability in SMEs, which is the motivation behind the current effort.

4.2.2 ISM Application

ISM has been used in understanding the interrelations among the different enablers, barriers, or factors in the literature. The paper by Kumar and Rahman (2017) suggested that the shareholders, supply chain partners, and push from customers are perilous for establishing a communally justifiable supply chain using ISM. Thamsatitdej et al. (2017) established three dimensions: Environmental, social, and economic, which need to be related in order to eco-design practices with the use of ISM and Matrices' Impacts Cruise's Multiplication Appliquée a UN Classement (MICMAC). Yadav and Singh (2020) have used ISM with a fuzzy analytic process to implement blockchain over the ongoing methods being used in the supply chain. Babu et al. (2020) used ISM to discover the interrelationship amidst supply chain risks, and the findings resulted in external risks having the most impact. Singh et al. (2020) suggested that the need for flexibility in the supply chain has improved. ISM is used for identifying the aspects of supply chain flexibility (SCF). Kumar and Rahman (2017) evinced that focal firms and sustainable supply chains are interrelated and bear enablers. The relationship among them is established using ISM. Digalwar et al. (2020) asked the experts and industrialists to evaluate the data in the field of lean-agile firms of India for ecological supply series practices using ISM. Literature has studied various researches that concentrate on social sustainability of various factors through ISM, but the ISM technique, in the case of social sustainability in

SMEs, has been overlooked, so our study aims to emphasize the study of communal sustainability in SMEs with the help of the ISM technique.

4.2.3 Economic, Innovation, and Leadership Dimensions

SMEs are influenced by innovation and economic aspects. The scope for SMEs is expanding because of the increase in manufacturing opportunities in developing countries. (Kumar & Subrahmanya, 2010). The opportunities that SMEs are provided with, keeping in mind the government schemes and aid, during times of recession can have diverging effects on the working of SMEs. (Soininen et al., 2012). When it comes to SMEs producing their output, the economic, social, and environmental feature of proactive CSR has a relationship with the vision, the stakeholders, and strategic proactivity (Torugsa et al., 2013). Furthermore, information and communication technologies (ICT) have great potential to achieve desired outcomes in SMEs (Tarutė & Gatautis, 2014). Internal innovation exists in SMEs, and there is a significant difference in the space of personal and organisational reasons that shape the origin of the modification amongst the two dimensions: Innovation and technology. (Martínez-Román & Romero, 2017). The pressure builds up in an enterprise due to its size, as it is small and comes with numerous resource-related barriers. To overcome this, internal knowledge and awareness of capabilities become paramount (Akhtar et al., 2015; Kumar et al., 2019). Export challenges should be overcome to further broaden the prospects of exporting (Paul et al., 2017). The economic enactment of SMEs indicates the turnover and development of the business. To strengthen financials, a relation among the social, environmental, and operational practices is documented (Malesios et al., 2018). Even though the financial contribution of SMEs keeps on snowballing everywhere, there are certain financial pullbacks due to inadequacy in leadership. Hence, the equilibrium between leadership and financial growth must be present. (Madanchian & Taherdoost, 2019). In the literature, it can be assumed that the main focus was related to the economic, innovation, and leadership dimensions of SMEs, and there is a lack of research done in the field of sustainability of SMEs. Hence, the main focus of our paper is on the sustainability dimension of SMEs.

This study is on business enablers of SMEs on social sustainability. SMEs pollute the environment and often hide this fact in order to protect themselves from facing legal issues. Moreover, water bodies are polluted because of the dumping of toxic waste into them. The study aims to concentrate on using legal and business enablers to protect eco-friendly environments and also produce desirable outcomes for SMEs. Furthermore, it emphasizes the challenges in using an appropriate legal structure for the social sustainability of SMEs.

4.3 RESEARCH METHODOLOGY

4.3.1 Identification of Factors

Legal and business dimension facilitators were identified in order to evaluate the social sustainability for the SMEs in India's northern region. The first step is to identify the enablers. The various enablers of the implementation of social sustainability were identified through an extensive literature survey. Databases, 'Google Scholar,' 'Science Direct,' and 'Web of Science' were searched using the keywords, 'sustainable supply chain management,' 'sustainable SMEs,' 'enablers for sustainability,' 'social sustainability,' and 'motivators for sustainability.' In these databases, we narrowed our search by assessing peer-reviewed journals in the field of sustainable supply chain and operations management. Through this process, we identified nine key enablers in achieving sustainability, which are subsequently listed in Table 4.2. The nine enablers identified are basically the elements of our ISM model, in which

TABLE 4.2
List of Legal and Business Enablers for SMEs in Attaining Social Sustainability

Notations	Sub-Enablers	Description	References
SE1	Employee training according to industry standards	The training of the employees is an important aspect of the organization. Training should be according to prescribed standards.	Gandhi et al. (2018)
SE2	Top management commitment	The top-level is the strategic level for the organisation. If the top-level is committed to their work, then it is always a plus for the enterprises.	Gandhi et al. (2018)
SE3	Use of information technologies to increase the efficiency of communication	To fill the communication gaps amidst the employees and departments, the use of technologies is necessary.	Kot (2018)
SE4	Building long-term relationships with channel partners	Channel partners are knowledgeable about market strategies and distribution channels, which can be effective in sealing products.	Kot (2018)
SE5	Use of 'just-in-time'	Inventory costs are reduced, thus, availing of products whenever needed.	Kot (2018)
SE6	Knowledge and information sharing between partners	Two-way communication and better partnership are promoted by sharing information and knowledge, therefore achieving organization goals.	Kot (2018)
SE7	Adherence to ISO 9000 guidelines	International standards for quality assurance are complied with in order to maintain the effective quality of the products.	Kot (2018)
SE8	Incorporation of customer feedback	Customer feedback helps and reshapes business operations.	Kot (2018)
SE9	Adherence to ILO standards	ILO is the standard for the labourers. It is important to adhere to these in order to set principles and rights at work.	Gandhi et al. (2018)

the interrelations are modelled. A panel comprised of experts, which include the secretary-general, the chairman from a reputable company, and the stakeholders from ventures, as well as academic scholars, referred to for additional understanding, the enablers in context with the legal and business dimensions, which enhance the socially sustainable working of the SMEs. Detailed explanations, personal interviews, and brainstorming sessions were conducted on the topics of social sustainability. A detailed description of the pressures was provided, and experts were approached for their opinion on defining the contextual relationship among the identified pressures.

4.3.2 ISM Methodology

ISM helps in defining the relationships among specific components which describe an issue (Warfield et al., 1974). The ISM process transforms ill-defined models into a clearly defined model (Darbari et al., 2018). It proposes the use of expert opinion to identify and form the relationship amidst different enablers and helps to form a structural relationship among them (Shankar et al., 2003). In this study, an ISM approach is used to identify the managerial implications and their implementation in SMEs (Shanker et al., 2019; Sindhwani et al., 2018; Sindhwani, Singh, Chopra, et al., 2019). Figure 4.1 demonstrates the various stages involved in the methodology.

After identifying the legal and business enablers with the help of the literature review and opinions from various industrialists and research scholars, the ISM methodology is then applied. Initially, the self-structural interaction matrix (SSIM) for enablers is developed. At this stage, the experts were asked to conduct a pairwise evaluation of each of the enablers. The matrix shows the relationship directly between the enablers of SMEs. The direction of the legal and business enablers is symbolized with these four symbols:

V – Enabler k will improve enabler l.
A – Enabler k will be improved by enabler l.
X – Enabler k and l will help achieve each other.
O – Enabler k and l are unrelated.

Based on the answers, a knowledge base was developed in the form of a table.

The reachability matrix is prepared with the help of a structural self-interaction matrix (SSIM) table. The SSIM table, which is represented with the help of the V, A, O, X symbols, is now converted into binary numbers 0 and 1 with the help of the following rules:

- If (k,l) in the SSIM is O, then (k,l) and (l,k) entries in the reachability matrix become 0 for both.
- If (k,l) in the SSIM is X, then (k,l) and (l,k) entries in the reachability matrix become 1 for both.
- If (k,l) in the SSIM is V, then (k,l) and (l,k) entries in the reachability matrix become 1 and 0, respectively.
- If (k,l) in the SSIM is A, then (k,l) and (l,k) entries in the reachability matrix become 0 and 1, respectively.

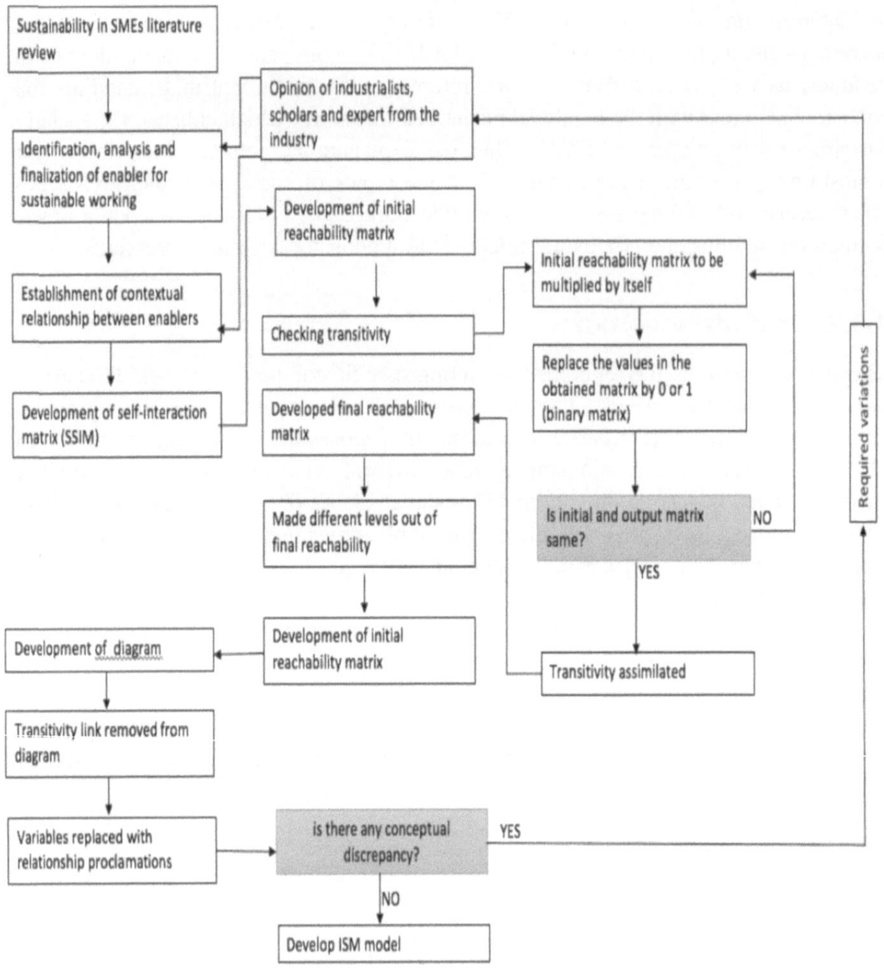

FIGURE 4.1 Flow Diagram of the ISM Methodology.

The initial reachability matrix obtained in the previous step has entries for only direct relationships. Based on the rule of transitivity, 'if A influences B, and B influences C', then it can be concluded that A influences C due to transitivity even though there is no direct relationship between A and C. Each possible transitive link creates an entry in the knowledge base, and the logical interpretation is renamed as transitive. Furthermore, the final reachability matrix is constructed by incorporating all of the transitivity in the initial reachability matrix. By the final reachability matrix, the reachability set (RS) and antecedent set (AS) are obtained for each of the legal and business enablers for social sustainability, and further the intersection of these sets results from the final reachability matrix (Sindhwani, Singh, Iqbal, et al., 2019). In the first iteration, the pressures with the same intersection and RS are designated at the topmost level (i.e. Level I), and the Level I elements are detached from the entire set. This iterative process is continued until each practice is assigned with levels. Then with the final reachability matrix ISM model is drawn. The graph is illustrated consistently with the levels obtained by level partitioning. The resulting graph is also called a digraph. A review of the ISM model is completed by checking inconsistencies and subsequently modifying the necessary changes.

4.4 CASE BACKGROUND

SMEs are non-subsidized and self-reliant firms that hire fewer employees. Section 7 of the Indian Penal Code (IPC) deals with the Micro, Small, and Medium Enterprises Development Act of 2006 (MSMED Act). The act provides a classification of MSMEs/ SMEs in two parts, the manufacturing and service enterprises. This classification is done based on investment size and the activities undertaken.

According to the SME Chamber of India, SMEs are the pillars that boost financial growth in the country. Other than the financial dimension, SMEs also prove to be a source of employment, innovation, exports, *etc*. They contribute highly to the economic growth of the nation.

The study by Mani et al. (2020), suggested that there is an effective relationship of social sustainability in SMEs, as large-scale enterprises are moving forward towards sustainability to increase efficiency and save energy. Sustainability in SMEs helps in sustaining in the changing world and balancing business from the economic, social, legal, and environmental impacts.

According to the UN Global Compact, social sustainability is about identifying the negative and positive effects of managing the business, on people. The kind of relationship that a business maintains with its stakeholders is crucial, and the business directly or indirectly affects its employees, workers, customers, and local communities, as listed in Table 4.3.

Despite government schemes like Skill Upgradation and Mahila Coir Yojana (2019), ISO 9000 guidelines, and ILO standards, the concerns of SMEs towards

TABLE 4.3
SSIM

	SE9	SE8	SE7	SE6	SE5	SE4	SE3	SE2	SE1
SE1	A	O	O	O	V	O	O	O	X
SE2	O	O	V	O	O	V	O	X	
SE3	O	V	O	V	O	V	X		
SE4	O	O	O	A	O	X			
SE5	O	A	A	O	X				
SE6	O	O	O	X					
SE7	V	A	X						
SE8	O	X							
SE9	X								

legal and business dimensions still pose a problem in terms of working in a sustainable method.

This brings us to the following objectives:

A. To help SMEs work in a sustainable manner while considering the legal and business environment.
B. To form an ISM model comprised of legal and business dimensions to collectively work with governmental guidelines.

4.4.1 Case Description

The SSIM table for the given dimensions is listed in Table 4.4. In the table, it states that employee training according to industry standards (SE1) will be more helpful in improving the use of 'just-in-time' (SE5), so there is a link between SE1 and SE5 which is consequently denoted by V. In SE1, i.e. employees training according to industry standards, and SE9, i.e. adherence to government local tax, SE9 helps to improve SE1, so relation A is used for 'just-in-time' to stand for SE9 and SE1. In the case of SE2, i.e. top management commitment, and SE9, i.e. adherence to government local tax, both are unrelated, so O is used to represent the relationship.

In the next step, a reachability matrix is formed by substituting V, A, O, and X with '0' and '1.' Here, SE4–6 holds the value A, which is substituted by 0, and therefore SE64 is updated to 1. Similarly, SE38 holds the value V, so it will be substituted as 1, and SE83 will be updated to 0. The cell with the value O will be simplified by 0, and the cell with value X will be simplified to 1. The final reachability matrix is prepared by incorporating the transitivities in the initial reachability matrix. The driving power consists of the enabler itself along with another enabler which it may impact. The dependence power also consists of the enabler itself and the enabler that may impact it.

TABLE 4.4
Initial Reachability Matrix

	SE1	SE2	SE3	SE4	SE5	SE6	SE7	SE8	SE9
SE1	1	0	0	0	1	0	0	0	0
SE2	0	1	0	1	0	0	1	0	0
SE3	0	0	1	1	0	1	0	1	0
SE4	0	0	0	1	0	0	0	0	0
SE5	0	0	0	0	1	0	0	0	0
SE6	0	0	0	1	0	1	0	0	0
SE7	0	0	0	0	1	0	1	0	1
SE8	0	0	0	0	1	0	1	1	0
SE9	1	0	0	0	0	0	0	0	1

The next table to be prepared is the level partitioning table. In this table, the first column represents the reachability set, which shows the legal and business enablers of its influences. Furthermore, the next column, i.e the antecedent set, shows the legal and business enablers influencing it. The intersection set, in which the intersection is equal to the RS, is labeled as Level 1. For example, SE4, SE5 is at level 1, in the table below. In order to obtain the next enabler, SE4 and SE5 are omitted from the table, and new reachability, antecedent, and intersection sets are identified. The second table contains SE1 and SE6, which are at Level 2. This continues until each enabler is defined by a level. A total number of six repetitions of the procedure were carried out. Level partitioning helps in the preparation of the ISM model and in the creation of the digraph.

4.5 RESULTS AND DISCUSSION

From the final reachability matrix, the diagram results are presented in the form of the ISM model, as disclosed in Figure 4.2. Furthermore, as compared with all of the

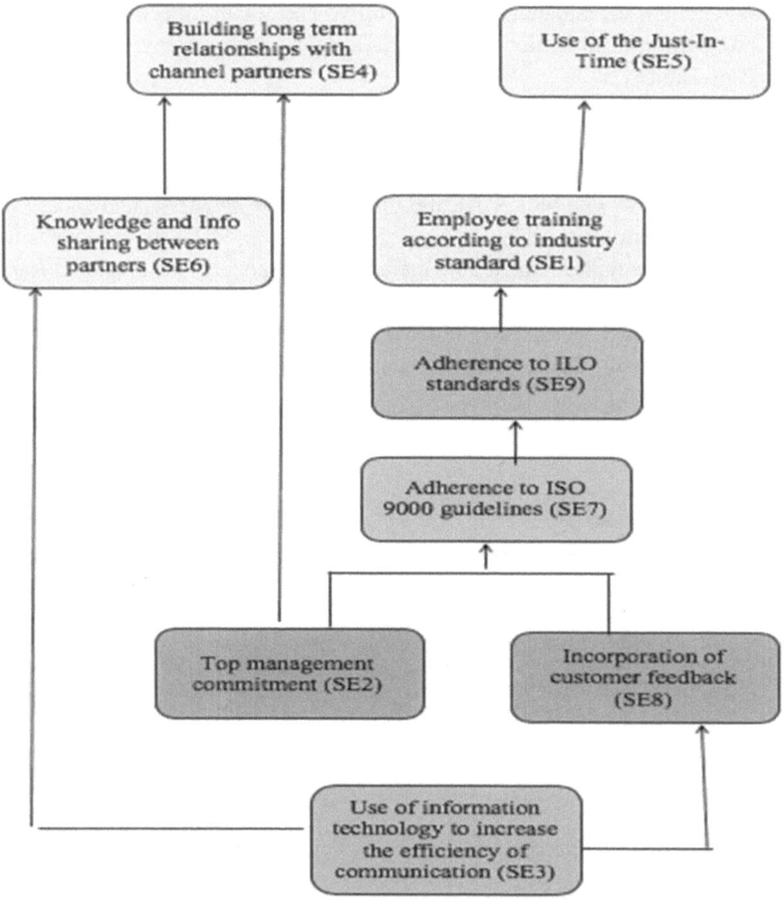

FIGURE 4.2 ISM Model.

legal and business enablers identified for the socially sustainable working of SMEs, the 'use of communication technology to increase the efficiency of communication' (SE3) is an independent enabler. It is also the most important enabler which will be most helpful to SMEs in achieving sustainability while keeping in mind the legal and business dimensions. The 'use of communication technology to increase the efficiency of communication' (SE3) will further lead to better top management commitment (SE2) and better incorporation of customer feedback (SE8). On the other hand, those on top, like 'building long-term relationships with channel partners' (SE4) and the 'use of just-in-time' (SE5) are the most dependent factors and bear the least importance, as compared with other dimensions, when it comes to achieving sustainability. The recent emphasis on sustainability had forced SMEs to consider its implementation on business operations. A greater understanding of the enablers, which will help in accomplishing the goal, has become indispensable. The use of information technology to increase the efficiency of communication (SE3) has become a vital component for organizations, proving that information communication technology (ICT) is the need of the hour. Hence, communication is the foundation element for enhancing sustainability. The social aspect of sustainability is even more difficult to achieve due to the human resources involved with it, as explained in Table 4.5

Thus, is it important that SMEs should work sustainably while considering the enablers in the legal and business dimensions in order to gain recognition and sustainably achieve desired results? With the help of Table 4.6, the MICMAC analysis is formed as depicted in Figure 4.3. All the nine legal and business enablers are positioned in four quadrants: Linkage, dependent, autonomous, and independent. In Quadrant I, i.e. the linkage quadrant, there is an absence of enablers. This is an indication of successful identification of legal and business dimensions, as their presence in Quadrant I would have created hindrances for SMEs in achieving sustainability. In Quadrant II, i.e. the dependant quadrant, there are top management

TABLE 4.5
Final Reachability Matrix

	SE1	SE2	SE3	SE4	SE5	SE6	SE7	SE8	SE9	Driver Power
SE1	1	0	0	0	1	0	0	0	0	2
SE2	0	1	0	1	1*	0	1	0	1*	5
SE3	0	0	1	1	1*	1	1*	1	0	6
SE4	0	0	0	1	0	0	0	0	0	1
SE5	0	0	0	0	1	0	0	0	0	1
SE6	0	0	0	1	0	1	0	0	0	2
SE7	1*	0	0	0	1	0	1	0	1	4
SE8	0	0	0	0	1	0	1	1	1*	4
SE9	1	0	0	0	1*	0	0	0	1	3
Dependence Power	3	1	1	4	7	2	4	2	4	

TABLE 4.6
I-Iteration

	Reachability Set	Antecedent Set	Intersection	Level
SE1	1, 5,	1, 7,9	1	2
SE2	2, 4, 5, 7, 9	2	2	5
SE3	3, 4. 5, 6, 7, 8	3	3	6
SE4	4	2, 3, 4, 6	4	1
SE5	5	1, 2, 3, 5, 7, 8, 9	5	1
SE6	4, 6	3, 6,	6	2
SE7	1, 5, 7, 9	2, 7, 8	7	4
SE8	5, 7, 8, 9	3, 8	8	5
SE9	1, 5, 9,	2, 7, 8, 9	9	3

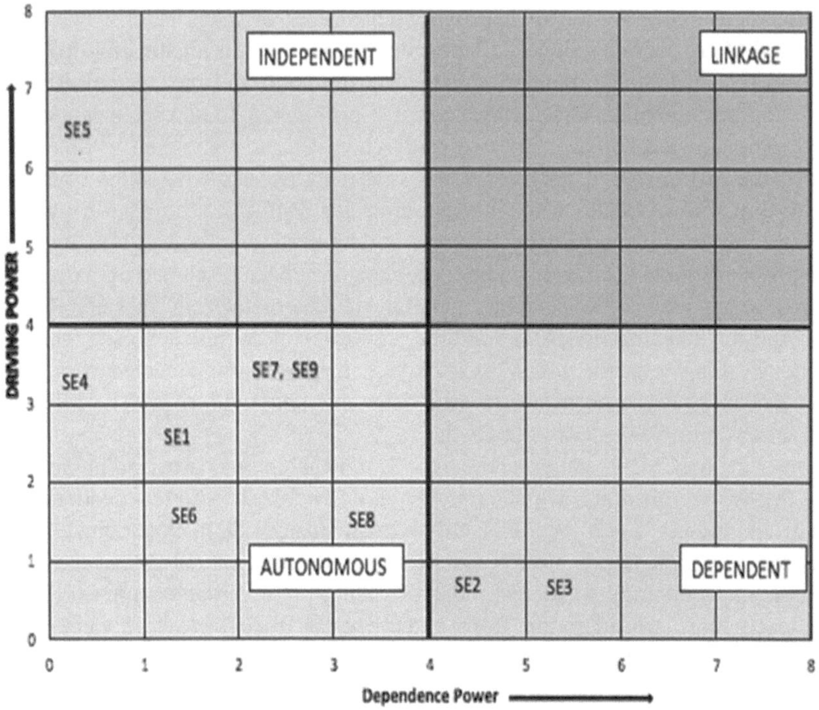

FIGURE 4.3 Classifications of Strategies (MICMAC).

commitment (SE2) and the use of IT to upsurge the efficiency of communication (SE3), which implies that these bear high dependent power. This means that even without directly focusing on these enablers, sustainability can be achieved because these are highly dependent on the other dimensions of legal and business development. The autonomous quadrant includes employee training according to industry standards (SE1), building long terms relationships with channel partners (SE4), knowledge and information sharing between partners (SE6), adherence to ISO 9000 guidelines (SE7), incorporation of customer feedback (SE8), and adherence to ILO standards (SE9). These enablers have the power to influence or be influenced. Therefore, whenever there is a change in any of the legal and business enablers, this will affect the other enablers and vice versa. The last quadrant, i.e. the independent quadrant, which includes the use of just-in-time (SE5), is the key strategy as it will not influence any other enabler despite changes in order to achieve sustainability.

4.6 CONCLUSION

Indian Small and Medium Enterprises (SMEs), given the sheer size and span they have, can play an exceptional part in strengthening ideas, inclusivity, and the enhancement of social sustainability in industrial development. They play a paramount part in amplifying the country's overall production networks, scale, and growth, with a contribution of approximately 6.11% of manufacturing GDP and around 24.6% of service sector GDP, despite the fact that it is challenging for SMEs to work in a sustainable manner, considering the legal and business dimensions. Achieving sustainability has always posed a problem for authorities working in SMEs. The aim is to answer the two research questions mentioned in this study: What is the part or role of SMEs in social sustainability, and what is the contextual relationship within social sustainability factors for SMEs?

This is the exact reason why the study addresses this problem. To overcome these problems, nine legal and business enablers have been discerned with the aid of experts, which include the secretary-general, the chairman from a reputable company, and the stakeholders from ventures, as well as academic scholars. The ISM technique, along with the MICMAC analysis, has been implemented in order to attain a result that can be of use to establishments and enterprises. The following observations and results were obtained:

As per the MICMAC analysis (Figure 4.3), no enabler was obtained in Quadrant I, i.e. the linkage quadrant, which depicts that no enabler has high dependence and high driving power. Apart from this, only two dimensions lie in Quadrant II, i.e. the dependent quadrant. These are top management commitment (SE2) and the use of information technology to increase the efficiency of communication (SE3), and these need more consideration from enterprises as they have high dependence power, which means that they are highly dependent on other enablers as well, and there exists scope for the improvement of these enablers. In the autonomous quadrant i.e. Quadrant III, the maximum number of enablers exist since the working and influencing power of these enablers on achieving sustainability by the SMEs is supreme. These include employee training according to industry standards (SE1), building long-term relationships with channel partners (SE4), knowledge and information sharing between partners (SE6), adherence to ISO 9000 guidelines (SE7), incorporation of customer feedback (SE8), and adherence to ILO standards (SE9). Additionally, the enabler lying in Quadrant IV, i.e. the independent quadrant, is the most important enabler, the use of just-in-time (SE5), in achieving sustainability.

The conclusion from the ISM model (Figure 4.2) is that the use of information technology to increase the efficiency of communication (SE3) is the most important legal and business enabler which will help SMEs in achieving sustainability. The two other vital strategies, as shown in Figure 4.2, are top management commitment (SE2) and the incorporation of feedback (SE8), which will also permit SMEs to work in a sustainable way in the legal and business dimensions. Furthermore, our study can be used by SMEs situated in northern India. The results can be changed by managers and researchers depending upon the enterprises and their dimensions. In the future, optimized models can be generated based on the results.

4.7 MANAGERIAL IMPLICATION

Small and Medium Enterprises (SMEs) situated in northern India are foreseen to work in a socially sustainable manner. These enterprises have driving forces that enable them to achieve the goal in the legal and business environments. In this light, the study has been done by implementing the key enablers required to achieve social sustainability in SMEs which provide dynamic managerial implications. Firstly, in these times, when the need to work sustainably has become greater, an understanding of the enablers which will help in achieving our goal has become quite indispensable. The use of information technology to increase the efficiency of communication (SE3) has become essential, as it has flattened the sustainable working of organizations. Information communication technology (ICT) is highly dynamic and has made rapid advancements. Enterprises using upgraded technology have worked in a more efficient way, and this has helped them to develop their business. ICT makes ends meet by creating transparent communication.

Top management commitment (SE2) is the second most important enabler. Top-level administration makes an action plan not only for resources, but also for formulating ways to implement sustainability in the everyday working conditions of the enterprise, as this subsequently highly powers the organisation's culture. The incorporation of customer feedback (SE8) is also the second most important enabler. Customers in today's world are active participants in various sustainable practices. Integrating their feedback would guide the organization to the right path. Adherence to ISO 9000 guidelines (SE7) is the third most important enabler. When followed, these guidelines lead to better productivity and lesser costs. This also enhances the quality of the products and improves their life cycle which, in turn, generates lesser waste.

REFERENCES

Akhtar, C.S., Ismail, K., Ndaliman, M.A., Hussain, J., & Haider, M. (2015). Can intellectual capital of SMEs help in their sustainability efforts. *Journal of Management Research*, *7*(2), 82.

Álvarez Jaramillo, J., Zartha Sossa, J.W., & Orozco Mendoza, G.L. (2019). Barriers to sustainability for small and medium enterprises in the framework of sustainable development—Literature review. *Business Strategy and the Environment*, *28*(4), 512–524.

Babu, H., Bhardwaj, P., & Agrawal, A.K. (2020). Analysis and model development of supply chain risk variables in the Indian manufacturing context. *International Journal of Business Continuity and Risk Management*, *10*(4), 307–329.

Battaglia, M., Testa, F., Bianchi, L., Iraldo, F., & Frey, M. (2014). Corporate social responsibility and competitiveness within SMEs of the fashion industry: Evidence from Italy and France. *Sustainability*, *6*(2), 872–893.

Centobelli, P., Cerchione, R., & Singh, R. (2019). The impact of leanness and innovativeness on environmental and financial performance: Insights from Indian SMEs. *International Journal of Production Economics*, *212*, 111–124.

Darbari, J.D., Agarwal, V., Sharma, R., & Jha, P.C. (2018). Analysis of impediments to sustainability in the food supply chain: An interpretive structural modeling approach. In P. Kapur, U. Kumar, & A. Verma (Eds)., *Quality, IT and Business Operations* (pp. 57–68). Springer, Singapore. https://doi.org/10.1007/978-981-10-5577-5_5

Das, M. (2019). Impact of Social and Environmental Practices on SME Business Performance. doi: 10.31124/advance.7749695.v1.

Digalwar, A., Raut, R.D., Yadav, V.S., Narkhede, B., Gardas, B.B., & Gotmare, A. (2020). Evaluation of critical constructs for measurement of sustainable supply chain practices in lean-agile firms of Indian origin: A hybrid ISM-ANP approach. *Business Strategy and the Environment*, *29*(3), 1575–1596.

Gandhi, N.S., Thanki, S.J., & Thakkar, J.J. (2018). Ranking of drivers for integrated lean-green manufacturing for Indian manufacturing SMEs. *Journal of Cleaner Production*, *171*, 675–689.

Halme, M., & Korpela, M. (2014). Responsible innovation toward sustainable development in small and medium-sized enterprises: A resource perspective. *Business Strategy and the Environment*, *23*(8), 547–566.

Johnson, M.P. (2017). Knowledge acquisition and development in sustainability-oriented small and medium-sized enterprises: Exploring the practices, capabilities and co-operation. *Journal of cleaner production*, *142*, 3769–3781.

Kot, S. (2018). Sustainable supply chain management in small and medium enterprises. *Sustainability*, *10*(4), 1143.

Kumar, D., & Rahman, Z. (2017). Analyzing enablers of sustainable supply chain: ISM and fuzzy AHP approach. *Journal of Modelling in Management*, *12*(3), 498–524.

Kumar, K., Dhillon, V.S., Singh, P.L., & Sindhwani, R. (2019). Modeling and analysis for barriers in healthcare services by ISM and MICMAC analysis. In M. Kumar, R. Pandey, & V. Kumar (Eds.), *Advances in Interdisciplinary Engineering*. Lecture Notes in Mechanical Engineering (pp. 501–510). Springer, Singapore. https://doi.org/10.1007/978-981-13-6577-5_47

Kumar, R., Kumar, V., & Singh, S. (2017a). Modeling and analysis on supply chain characteristics using ISM technique. *Apeejay Journal of Management and Technology*, *12*(1 & 2).

Kumar, R., Kumar, V., & Singh, S. (2017b). Modeling and analyzing the impact of lean principles on organizational performance using ISM approach. *Journal of Project Management*, *2*, 37–50.

Kumar, R., Kumar, V., & Singh, S. (2017c). Work culture enablers: Hierarchical design for effectiveness & efficiency. *International Journal of Lean Enterprise Research (IJLER)*, *2*(3), 189–201.

Kumar, R.S., & Subrahmanya, M.B. (2010). Influence of subcontracting on innovation and economic performance of SMEs in Indian automobile industry. *Technovation*, *30*(11–12), 558–569.

Madanchian, M., & Taherdoost, H. (2019). Assessment of leadership effectiveness dimensions in small & medium enterprises (SMEs). *Procedia Manufacturing*, *32*, 1035–1042

Malesios, C., Skouloudis, A., Dey, P.K., Abdelaziz, F.B., Kantartzis, A., & Evangelinos, K. (2018). The impact of SME sustainability practices and performance on economic growth from a managerial perspective: Some modeling considerations and empirical analysis results. *Business Strategy and the Environment*, *27*(7), 960–972.

Martínez-Román, J.A., & Romero, I. (2017). Determinants of innovativeness in SMEs: Disentangling core innovation and technology adoption capabilities. *Review of Managerial Science*, *11*(3), 543–569.

Meath, C., Linnenluecke, M., & Griffiths, A. (2016). Barriers and motivators to the adoption of energy savings measures for small-and medium-sized enterprises (SMEs): The case of the ClimateSmart Business Cluster program. *Journal of Cleaner Production*, *112*, 3597–3604.

Mittal, V.K., Sindhwani, R., Shekhar, H., & Singh, P.L., (2019). Fuzzy AHP model for challenges to thermal power plant establishment in India. *International Journal of Operational Research*, *34*(4), 562–581.

Narula, R. (2004). R&D collaboration by SMEs: New opportunities and limitations in the face of globalisation. *Technovation*, *24*(2), 153–161.

Omerzel, D.G., & Antončič, B. (2008). Critical entrepreneur knowledge dimensions for the SME performance. *Industrial Management & Data Systems*, *108*(9), 1182–1199.

Panizzolo, R., Garengo, P., Sharma, M.K., & Gore, A. (2012). Lean manufacturing in developing countries: Evidence from Indian SMEs. *Production Planning & Control*, *23*(10–11), 769–788.

Paul, J., Parthasarathy, S., & Gupta, P. (2017). Exporting challenges of SMEs: A review and future research agenda. *Journal of world business*, *52*(3), 327–342.

Rizos, V., Behrens, A., Van der Gaast, W., Hofman, E., Ioannou, A., Kafyeke, T., ... & Topi, C. (2016). Implementation of circular economy business models by small and medium-sized enterprises (SMEs): Barriers and enablers. *Sustainability*, *8*(11), 1212.

Shanker, K., Shankar, R., & Sindhwani, R. (Eds.). (2019). *Advances in industrial and production engineering: Select proceedings of FLAME 2018*. Springer.

Shankar, R., Narain, R., & Agarwal, A. (2003). An interpretive structural modeling of knowledge management in engineering industries. *Journal of Advances in Management Research*, *1*(1), 28–40.

Sindhwani, R., Mittal, V.K., Singh, P.L., Kalsariya, V., & Salroo, F. (2018). Modelling and analysis of energy efficiency drivers by fuzzy ISM and fuzzy MICMAC approach. *International Journal of Productivity and Quality Management*, *25*(2), 225–244.

Sindhwani, R., Singh, P.L., Chopra, R., Sharma, K., Basu, A., Prajapati, D.K., & Malhotra, V. (2019). Agility evaluation in the rolling industry: A case study. In K. Shanker, R. Shankar, & R. Sindhwani (Eds.), *Advances in Industrial and Production Engineering*. Lecture Notes in Mechanical Engineering (pp. 753–770). Springer, Singapore. https://doi.org/10.1007/978-981-13-6412-9_70

Sindhwani, R., Singh, P.L., Iqbal, A., Prajapati, D.K., & Mittal, V.K. (2019). Modeling and analysis of factors influencing agility in healthcare organizations: An ISM approach. In K. Shanker, R. Shankar, & R. Sindhwani (Eds.), *Advances in industrial and production engineering*. Lecture Notes in Mechanical Engineering (pp. 683–696). Springer, Singapore.

Singh, R.K., Garg, S.K., & Deshmukh, S.G. (2007). Interpretive structural modelling of factors for improving competitiveness of SMEs. *International Journal of Productivity and Quality Management*, *2*(4), 423–440.

Singh, R.K., Joshi, S., & Sharna, M. (2020). Modelling supply chain flexibility in the Indian Personal Hygiene Industry: An ISM-Fuzzy MICMAC approach. *Global Business Review*, doi: 10.1177/0972150920923075.

Soininen, J., Puumalainen, K., Sjögrén, H., & Syrjä, P. (2012). The impact of global economic crisis on SMEs. *Management Research Review*, *35*(10), 927–944.

Storey, D.J. (2006). Evaluating SME policies and programmes: Technical and political dimensions. In *The Oxford handbook of entrepreneurship*.

Stubblefield Loucks, E., Martens, M.L., & Cho, C.H. (2010). Engaging small- and medium-sized businesses in sustainability. *Sustainability Accounting, Management and Policy Journal*, *1*(1), 178–200.

Tarutė, A., & Gatautis, R. (2014). ICT impact on SMEs performance. *Procedia-social and behavioral Sciences*, *110*, 1218–1225.

Thamsatitdej, P., Boon-Itt, S., Samaranayake, P., Wannakarn, M., & Laosirihongthong, T. (2017). Eco-design practices towards sustainable supply chain management: Interpretive structural modelling (ISM) approach. *International Journal of Sustainable Engineering*, *10*(6), 326–337.

Torugsa, N.A., O'Donohue, W., & Hecker, R. (2013). Proactive CSR: An empirical analysis of the role of its economic, social and environmental dimensions on the association between capabilities and performance. *Journal of Business Ethics*, *115*(2), 383–402.

Yadav, S., & Singh, S.P. (2020). An integrated fuzzy-ANP and fuzzy-ISM approach using blockchain for sustainable supply chain. *Journal of Enterprise Information Management*, 34(1), 54–78.

Warfield, J.N. (1974). Towards Interpretation of complex structural models. IEEE Transactions: System, Man and Cybernetics, *SMC-4*(5), 405–417.

5 Application of the MCDM Technique for an Astute Decision-Making Process in the Renewable Energy Sector

Eshan Bajal, Alakananda Chakraborty, Muskan Jindal, and Shilpi Sharma

CONTENTS

5.1 Introduction 83
5.2 Multi-Criteria Decision-Making Methods in Energy 86
5.3 Application of MCDM Methods for Energy Selection and Decision-Making Policies 87
5.3.1 Analytic Hierarchy Process (AHP) 88
5.3.2 Analytic Network Process (ANP) 89
5.3.3 Technique for Order of Preference by Similarity to Ideal Solution (TOPSIS) 90
5.3.4 ELimination Et Choix Traduisant la REalité (ELECTRE) 90
5.3.5 Preference Ranking Organization Method for Enrichment of Evaluations (PROMETHEE) 91
5.3.6 Combined Method Style Approach 92
5.4 Scrutiny Analysis of Papers about Energy Management in Policy-Making 93
5.5 Conclusion 95
References 101

5.1 INTRODUCTION

In this day and age, the constant availability of energy in the form of electricity has become relatively substantial that it can be labelled as a basic necessity for the sustenance of life. This holds true for all aspects of society, whether it be social, political, or economic. After the industrialization and wide-scale automation of manual labour, the demand for energy has experienced an exponential increase worldwide. Countries with an abundant supply of natural resources that can be used to generate electricity leverage this to gain a higher position in the world economy. On account of the limited availability of non-renewable resources, such as coal and crude oil, as well as the dangers associated with nuclear sources, renewable energy is gaining more and more attention every day. Not only are these easy to obtain with little capital, but these also do not have irreversible ill effects on the environment. As a result, this has become a field of considerable importance to researchers. However, multiple options for renewable energy are available for a place at any given time. The choice that would yield the greatest benefit in terms of long-term reliability, maintenance cost, and socio-political change is not discernible from the get-go. Hence, scientific and standardized techniques are used to analyze the various factors in order to arrive at the best possible outcome. One such method that is both easy to use and produces reliable results consistently across multiple different needs is multi-criteria decision-making (MCDM).

In multi-criteria decision-making, various parametres are analyzed in order to select an option amongst many that will provide the most output in accordance with the set criteria (Kumar et al., 2017a; Shanker et al., 2019; Sindhwani & Malhotra 2017; Sindhwani, Singh, Iqbal, et al., 2019). This tool has been in extensive use in the scientific and software community for more than 20 years; recently, its use has begun in other sectors, such as the energy, medical science, and geography industries. In case the data is too unfiltered for use in the classical MCDM method, fuzzy MCDM can be applied in order to gain more accurate results (Mittal et al., 2018; Sindhwani et al., 2018; Sindhwani, Singh, Chopra, et al., 2019). Different uses in energy application fields can be power source selection, long-term cost estimation, political and economic viability prediction, policy determination, and implemented labour evaluation. Various contemporary literature already discusses the application of MCDM to energy management, which deserves perusal. The crux of these studies is compiled as follows:

Assessment of sustainability with a grey-based model was suggested by (Arce et al., 2015). Their work was focused on AHP showing its benefits in relational analysis among energy factors. A different model system consisting of the outranking model, value measurement model, and goal model was proposed by (Kumar, Sah, et al., 2017) in the field of renewable energy complemented by a review of relevant related papers. A review in energy planning until 2004 was presented in (Pohekar & Ramachandran, 2004). Research similar to Kumar 2017) was done in (Strantzali & Aravossis 2016) applying MCDM in real-life problems. A different review study (Suganthi et al., 2015) delved into the criteria and weights associated with those in the problems. Implementation of fuzzy MCDM to determine alternative sources of energy while satisfying present requirements was done in (Kilic & Kaya 2015).

In the process of reviewing the studies, it was observed that research generally focuses on one aspect of MCDM, be it criteria selection, analysis, fuzzy implementation, or sustainability evaluation (Kumar et al., 2019, Mittal et al., 2019). In this chapter, a comprehensive and multi-field-encompassing review and analysis have been done with suitable modifications to make it relevant in the near future. Research work dealing with both renewable and non-renewable energy was explored, involving papers from 1988 to 2019, making this the most comprehensive piece of work to date. Furthermore, all evaluation criteria used in the papers cited are properly explained. Statistical work done in other studies has also been assimilated for ease of access in a single piece of literature, as shown in Figure 5.1–5.4. This study reveals the current trends and demands on-going in the field of MCDM, thus making it easy to recognize the direction this field is going in the future.

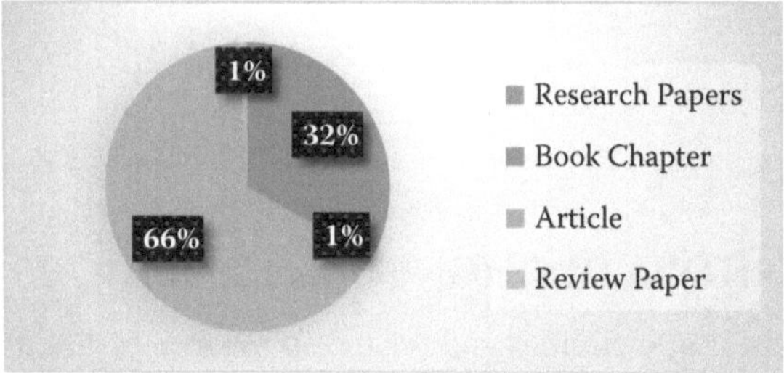

FIGURE 5.1 The Relative Quantity of Papers Reviewed by Type.

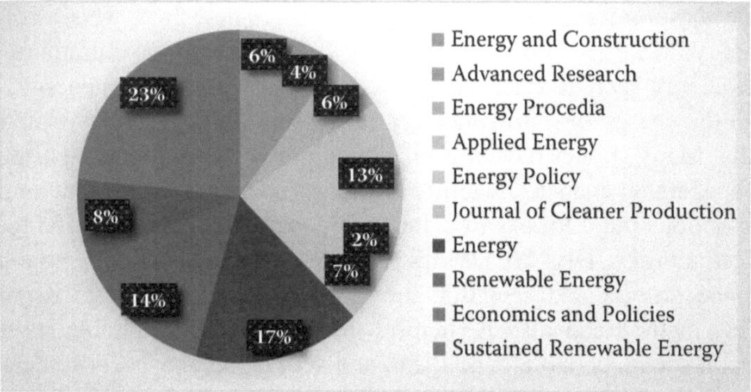

FIGURE 5.2 Segregation of Papers on the Basis of Fields of Research.

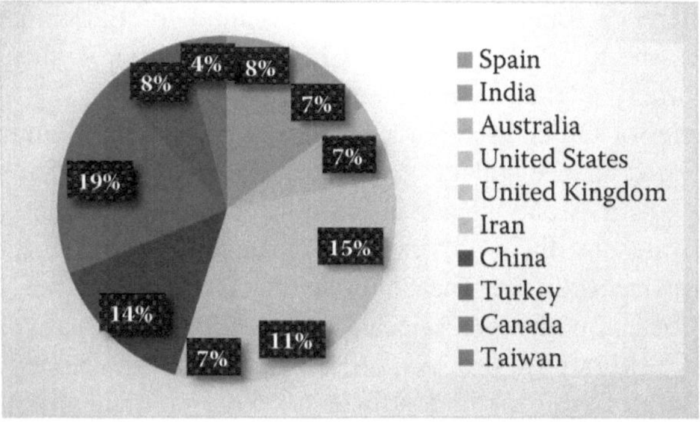

FIGURE 5.3 Classification of Papers on the Basis of Country of Implementation.

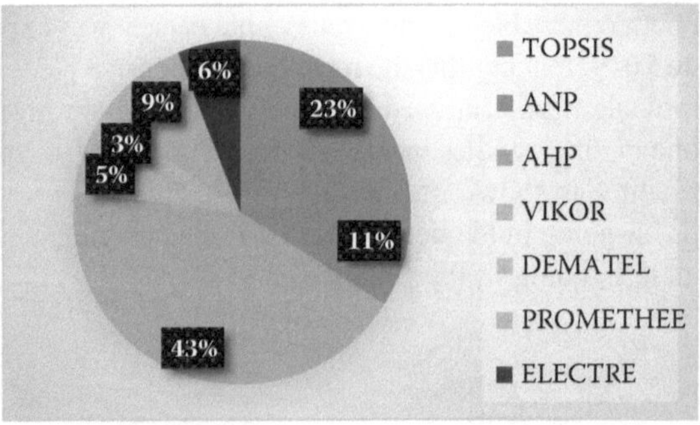

FIGURE 5.4 Segregation of Reviewed Papers on the Basis of MCDM Technique Used.

5.2 MULTI-CRITERIA DECISION-MAKING METHODS IN ENERGY

Energy is required for all activities and for the sustenance of life itself. The source of all energy can be traced back to the sun. The energy from coal, natural gas, and petroleum is predominantly the energy from the sun that was stored in plants and animals millions of years ago. Hence, these are non-renewable sources, meaning that is once these are exhausted, these cannot be replenished. This also includes nuclear energy, as this method is only viable as long as we have uranium to mine. Renewable energy such as solar energy, wind energy, geothermal energy, biomass energy, and the energy of a flowing river are all renewable, as these would not run out naturally. Many studies have associated with these types of energy usage, along with the management and sustainability issues. As energy is in constant demand, the developing nations put a lot of effort into using analytical and scientific tools to be able to select appropriately. MCDM has experienced extensive use in the aspect of evaluating the criteria and selecting the alternative that fits best (Kumar et al. 2017b). This method can also be helpful in determining a suitable solution that conforms with social, economic, political, and geographical aspects associated with the construction of an energy production facility. The main methods of MCDM applied are the Technique for Order Preference by Similarity to Ideal Solutions (TOPSIS), the Analytic Hierarchy Process (AHP), the Analytic Network Process (ANP), the Preference Ranking Organization Method for Enrichment Evaluation (PROMETHEE), and the Elimination and Choice Translating Reality (ELECTRE). These shall be briefly explained in the next section.

The Technique for Order Preference by Similarity to Ideal Solution (TOPSIS), originally suggested in (Şengül et al., 2015), is a technique that measures the distance of the positive and negative solutions from the current solution. The alternatives are rated by their proximity or distance from the ideal positive and negative solution respectively. The solution that is closest to the ideal positive without any major compromise is selected as the suitable alternative.

The Analytic Hierarchy Process (AHP) was first developed by Thomas (Cheng & Li, 2001). In this method, a three-level hierarchical structure is formed with the alternatives at the lowest level, the criteria at the second level, and the goal at the top. The alternative with the best weight according to features selected for use. Since it relishes simplicity, the process gives a higher level of focus to the individual factors. This, however, means that the system performance declines when higher-level mathematics is involved, as it cannot account for such complex relations. Even factors that are independent of parametres can lead to a discrepancy in results; all of these factors limit the cases where they can be applied effectively.

The Analytic Network Process (ANP) takes a more mathematical approach with matrices arithmetic used for all calculations. These systems work in conjugation with a feedback loop to evaluate all of the variables and sub-cluster interactions, thus ensuring better performance.

The Preference Ranking Organization Method for Enrichment Evaluation (PROMETHEE) incepted in the latter part of the 20th century is an outranking system (Behzadian et al. 2010). There are two variants of this method in general use: PROMETHEE I and PROMETHEE II. The former performs partial ranking, while the latter enforces a complete ranking of the available solutions. Being a highly mathematical-based formula, this too is of great scientific use because of its high versatility. The method compares each pair of solutions, discarding the one with the lower score until only one solution remains. A preference function ranging from 0 to 1 is calculated based on the usefulness of each parametre. Six preference functions are generally utilized, namely, linear, Gaussian, levelled, U-shaped, and V-shaped. Two powerful features, elimination and Choice Translation Reality, provide the gateway to analysis among different classes of alternative options by omitting the differences while making small adjustments to account for any discrepancy. In addition, these features allow for a wider range of applications, making it a lucrative choice over the other methods in the energy management domain.

Finally, similar to PROMETHEE, Elimination and Choice Translating Reality (ELECTRE) is the second MCDM technique that possesses the capacity to compare different options from different classes. It does so by an amalgamation of qualitative and quantitative analysis of available alternatives. Also similar to PROMETHEE, its internal working is comprised of systematic pair-wise matrix comparison with a strong focus on dominance relationship. The different variants of this are constituted of ELECTRE I, II, III, and IV, each serving a suitable purpose (Akash et al., 1999).

There are also several other methods in the vast ocean of MCDM that are currently discussed in detail. Some notable techniques are the Decision-Making Trial and Evaluation Laboratory (DEMATEL), Vise Kriterijumska Optimizacija I Kompromisno Resenje (VIKOR), Data Envelopment Analysis (DEA), Complex Proportional Assessment (COPRAS), Weighted Aggregated Sum Product Assessment (WASPAS), and Simple Additive Weighting (SAW). All of these techniques have certain implementations in energy-based problems.

5.3 APPLICATION OF MCDM METHODS FOR ENERGY SELECTION AND DECISION-MAKING POLICIES

Multi-Criteria Decision-Making is an efficient technique enabling the estimation of complications related to making decisions by applying analytic methods in procedures of making decisions within a wide range of criteria. MCDM techniques, their utilization, and their application in problems related to energy have become an important subject of various research works done to date. This section of the study presents a systematic and chronological literature review on the types of MCDM techniques and the studies based on their utilization and application in solving problems on planning energy.

5.3.1 Analytic Hierarchy Process (AHP)

This method was applied by (Kagazyo et al. 1997) where resource, social criteria, and technological issues were assessed for the prioritization of energy projects to determine that the electricity supply-system in Japan consisted of the maximum amount of power generation from the combustion of coal. The AHP method was used by (Akash et al. 1999) to compare among various energy generation alternatives, to conclude that the best substitutes were hydropower energy, solar energy, and wind energy, while nuclear energy and fossil fuels proved to be the worst for electricity generation. (Xiaohua & Zhenmin 2002) developed an index system to evaluate the relationship between sustainable development and energy in remote locations by using AHP to calculate the weight of each index. It was also applied at a university campus in Eskeşehir, Turkey by (Aras et al. 2004), to determine the most appropriate site for placing a station for wind observation where the most significant criteria were selected as the security and the regional topography. A methodology for making decisions on the basis of the AHP process was proposed by Kablan (2004) to prioritize the policy tools related to the conservation of energy, from a Jordan-based case study. It was observed that for the conservation of energy resources, the most applicable policies indicated were namely, 'law-making and execution,' 'fiscal incentives,' and 'qualification, education, and training.' A methodology was developed by (Wang et al. 2008) on the assessment of grey incidence for distributed systems on triple generation, which involved procedures of entropy information and AHP, along with weighting linear combinations. Five alternatives were evaluated using this model, each obtaining distinct results for assessments.

(Talaei et al. 2014) used this method in Iran to rank various novel technologies related to low-carbon energy based on various criteria. They concluded that oil and gas have the highest priority in technological usage for electricity and transport. A decision-making framework was improved upon by (Rosso et al. 2014) with the application of the AHP method to provide an effective assessment of hydropower projects for the construction of hydropower plants in mountain areas. They integrated multi-criteria evaluation and stakeholder analysis to determine from the study that the first project is the most applicable alternative. A model was proposed by (Bojesen et al. 2015) for planning and making decisions in which they applied multi-criteria estimation from spatial perspective and AHP to calculate priorities of the criteria, thereby defining and ranking the locations in Denmark that are convenient for the production of biogas. An AHP-based MCDM procedure was developed by (Al Garni et al., 2016) to evaluate five types of renewable energy resources in environmental, socio-political, technical, and economic criteria and ranked them accordingly at the end of the research. AHP was also applied by (Shirgholami et al., 2016) to define an evaluation criterion for assessing wind turbines in order to develop a wind farm. Four wind turbines were prioritized, and the best option that was selected was T-1. The method was again used to estimate the suitability of wind port alternatives for the purpose of the three lifecycle phases of wind projects positioned offshore, as undertaken by (Akbari et al., 2017). Two main important criteria were chosen, and Sheerness Port was concluded as the best alternative at the end of the research study. The deployment of solar energy in the Indian region was estimated by (Sindhu et al., 2017) by conducting a Strength-Weakness-Opportunities-Challenges (SWOC) analysis, prioritizing the variables by utilizing the AHP technique. In Paraguay, four energy policy alternatives were prioritized and evaluated by (Blanco et al., 2017) using an MCDM model based on the AHP under five main criteria in order to determine that the most suitable course of action would be the development of industrial zones of limited size. Integration of the MCDM model and AHP was proposed by (Haddad et al., 2017) for the purpose of the renewable energy resources in Algeria to be evaluated and the alternatives be ranked accordingly. In Jordan, different scenarios were developed, and the alternatives of energy resources were ranked appropriately by (Malkawi & Azizi, 2017) using AHP. Their results conclude the conventional sources to be the more suitable alternative for the area. (Sagbansua & Balo, 2017) evaluated six alternatives, various main criteria, and sub-criteria in order to rank the best to be the wind turbine.

5.3.2 Analytic Network Process (ANP)

For the purpose of evaluating energy options in the region of Turkey, (Ulutaş, 2005) used the ANP technique to determine that the most suitable alternative of energy resource is biomass. (Erdoğmuş et al., 2006) evaluated fuel alternatives for the use of residential heating by applying the ANP technique with a group decision-making approach and BOCR for determining the criteria weights. They concluded natural gas is the best alternative. The method was the basis of an MCDM model proposed by (Köne & Büke, 2007) to determine the most suitable mixture of energy resources in Turkey for the purpose of the generation of electric power. The BOCR method was applied to identify the priority weights of the criteria, and it was observed that the highest priority was ascribed to hydropower energy, while the lowest priority was given to energy generated from oil. This MCDM technique was also used in the study conducted by Önüt et al. 2008, Afsordegan 2014) for indicating which fuel resources were the most applicable for industries dealing with manufacturing, and they observed that the most widely used alternative was credited to electric power, while the least was fuel-oil energy. ANP was used to calculate the criteria weights for a methodology consisting of an amalgamation of BOCR and a model balancing scorecard, as presented by (Shiue & Lin, 2012), indicating that the most applicable classification of recycling strategy is in the industries dealing with solar power. The best strategy determined was 'in-house,' at the end of the study. (Kabak & Dağdeviren, 2014) combined the ANP and BOCR methods to present an MCDM model which was used to classify numerous criteria, in order to assess five alternatives of energy resources. The study concluded that energy generated from hydropower is the finest energy alternative.

5.3.3 Technique for Order of Preference by Similarity to Ideal Solution (TOPSIS)

Boran et al. (2013) analyzed multiple criteria by applying the TOPSIS technique for a comparative analysis of the contemporary sources of energy with nuclear power for the generation of electricity in the region of Turkey. These alternatives were analyzed with respect to various criteria, such as carbon dioxide emission, efficiency, acceptability, and total cost generation on the basis of various scenarios, and it was concluded that hydropower is the best alternative, followed by nuclear power, as evaluated for three scenarios. TOPSIS was utilized by (Brand & Missaoui, 2014) to evaluate five different energy scenarios under economic, security of supply, socio-economic, and ecological criteria to determine that the most acceptable scenario selected for Tunisia was the diversification of renewable energy sources. (Afsordegan et al., 2014) compared qualitative TOPSIS and a non-compensatory outranking MCDM method to present a multi-criteria analysis framework, for the selection of wind farm locations in Catalonia. At the end of the study, it was observed that both of these methods have similar performance qualifications. An extended version of the TOPSIS method, along with the weighted sum processes, was utilized by (Kablan, 2004) to present an analysis of multiple energy-planning criteria for risk-prioritizing in the development of projects involving tidal energy. It was observed that prioritizing risks would have a conflict between academy and industry. (Alidrisi & Al-Sasi, 2017), with a TOPSIS-based MCDM approach, intended to rank the countries that fall under G20 with respect to power generation from various sources of renewable energy. It was concluded that Germany and France were ranked the first for two distinct situations assessing the respective safe and unsafe energy resource as the nuclear power.

5.3.4 ELimination Et Choix Traduisant la REalité (ELECTRE)

A utility theory was utilized by (Roy & Bouyssou, 1986) to examine the MCDM approaches and the ranking distinction on practical grounds. The outranking procedure of ELECTRE III was applied for the selection of plant locations producing nuclear energy, and it was analyzed how situations related to real-life could be the target for adopting outranking MCDM techniques. (Georgopoulou et al., 1997) applied the tri-version of the same technique for the selection of policies dealing with renewable energy in the Greek islands to indicate the requirement of evaluation in various criteria, except in terms of minimization of expenditure for problems related to policy-making related to energy on a local basis. For the planning of energy resources, the study estimated eight strategies. Out of which, two were identified as the most suitable. The technique was also applied, along with the utilization of a case study in Sardinia island, to rank three different scenarios and evaluate a plan of action dealing with the combination of renewable energy resources at a local level, as proposed in an MCDM approach by (Beccali et al., 2003). The most suitable alternative scenario chosen was the reduction of the use of fossil fuels.

(Boemi et al., 2010) intended to indicate the renewable energy alternatives that were possible to be applied in an insular system for the production of electric power by applying the ELECTRE III method. The alternatives were assessed on the basis of technical, social, environmental, and financial criteria, and the results suggested that, while photovoltaic was inefficient and too expensive, wind power was categorized as the best alternative for providing electricity. The ELECTRE TRI technique was utilized by (Karakosta et al. 2009) to rank the sustainable energy sources for the generation of electricity in a few selected countries. (Catalina et al., 2011) used the ELECTRE III procedure for multisource of systems in the selection and design of appropriate alternatives and obtained reasonable results by applying a case study to this developed approach.

5.3.5 Preference Ranking Organization Method for Enrichment of Evaluations (PROMETHEE)

The outranking version of PROMETHEE II was used for the analysis and categorization of the alternatives of renewable energy resources by (Georgopoulou et al., 1998), and Greece was selected for a real-life case study, where renewable energy resources were exploited, for this proposed framework. In order to favour a user-friendly evaluation of multiple criteria in the projects involving renewable energy resources, a methodology for collective decision-making in pragmatic scenarios was proposed by (Haralambopoulos & Polatidis, 2003), in which the PROMETHEE II outranking approach was included. This developed a union between various users and their conflicting criteria. The method was also applied by (Topcu & Ulengin, 2004) for the identification of the most favourable alternative for electric power in the region of Turkey, and wind energy was determined as the solution. PROMETHEE was the basis of an MCDM approach proposed by (Madlener & Stagl, 2005) for the creation and design of instruments for the policies of renewable energy. It was observed that a reliable and efficient analysis of the lifecycle of renewable energy resources was provided by the proposed framework. For the generation of electric power in Greece, sustainable technologies were assessed by (Doukas et al., 2006) in implementing the PROMETHEE II technique. Based on four scenarios, criteria weights were provided, and the alternatives were analyzed to identify that the most favourable ones were wind energy, biomass energy, and lignite.

Again, for its Trozinia district, an MCDM framework consisting of an amalgamation of PROMETHEE and methods involving acceptance of society was developed by (Polatidis & Haralambopulos, 2007) to formulate the integration of renewable energy resources and to determine the priority of various alternatives. (Kowalski et al., 2009) utilized PROMETHEE as an MCDM technique for the determination of the alternatives of energy for Austria in the future, along with their prioritization. For this purpose, various criteria for sustainability and scenarios of renewable energy were evaluated. For the planning of renewable energy alternatives in remote locations, a methodology was prepared by (Terrados et al., 2009) in which they used PROMETHEE to quantify and rank the energy resources. (Tsoutos et al., 2009) selected various criteria for planning sustainable energy alternatives in Crete Island. For each alternative, an evaluation matrix was developed for unique users, and PROMETHEE was applied for final analysis. The technique was also utilized for the analysis of multiple criteria in the assessment of technologies related to solar thermal power, which was conducted by (Cavallaro, 2009). For heating of the district of Vancouver in Canada, the alternatives of energy resources were prioritized and evaluated by (Ghafghazi et al., 2010) by utilizing the technique of PROMETHEE. They selected various criteria, such as emission of GHG, technology maturity, emissions of particulate matter, and traffic load, and on this basis, they evaluated the accessible alternatives of energy to obtain definite results to distinct scenarios.

Chatzimouratidis and Pilavachi (2012) took 12 criteria under consideration and evaluated 10 power plants. They collected results for each of the scenarios generated by providing weights to each criterion and concluded that biomass, photovoltaic power, and wind are the most favourable power plants for most of the scenarios. To achieve sustainability in Scotland on a nation-wide scale, 11 renewable resource technologies were prioritized. Each criterion was provided equal weights, and ultimately, photovoltaic technology was selected as the most suitable alternative. In a village in Germany, the projects dealing with bioenergy at the regional level were evaluated by applying PROMETHEE as the basis for an MCDM framework, as developed by (Lerche et al., 2019). The best alternative, determined by prioritizing four energy options against various criteria, was the biogas plant. In Turkey, (Özkale et al., 2017) utilized the PROMETHEE method to assess five alternatives of power plants dealing with renewable energy against various criteria with equal priority weights. The hydroelectric power plant was declared as the most favourable alternative.

5.3.6 Combined Method Style Approach

In the works of (Kim et al., 1999), reprocessing of nuclear waste was accepted as the best alternative for fuel cycles in Korea. (Goletsis et al., 2003) did similar work in the American energy industry. The combined use of AHP amalgamated with SIM provided feasibility in an American setting, as shown by (Nigim et al., 2004). Their work categorized non-renewable sources in the diarchy of solar, followed by thermal, followed by wind energy. (Lee et al., 2008) proved that the most efficient energy sources were the non-carbon-dioxide-based sources used in the industry utilizing a DEA-based approach. (Ren et al., 2009) considered different scenarios in rural Japan. This data was fed to a linear algebra-based PROMETHEE system to fix the residual energy system. (Wang et al., 2008) provided a similar evaluation in the Chinese subcontinent. AHP-based energy evaluation of coal, natural gas, and crude oil concluded renewable energy as significantly more efficient than both coal and natural gas. (Al-Yahyai et al., 2013) present a review of energy services in Oman. The review also studies the economical, technical, and environmental factors that play a role in the region's power selection. (Daim et al., 2010) found alternatives to windmills by using the fuzzy Delphi method, followed by condensation matrix. This research concluded that compressed air energy storage is the superlative alternative. (Balin et al., 2012) conducted an MCDM comparison in the electricity production techniques in Turkey. An interesting finding was that ELECTRE was not a suitable method for outranking-based systems. Ultimately, PROMETHEE selected wind sources as the best renewable alternative. (Streimikiene et al., 2018) embarked on the quest to discover the best energy alternative amongst all sources. The calculations and evaluations were done with TOPSIS and MOORA, which showed hydroelectricity as the most efficient, and solar energy as highly maintainable over a long period of time. Certain remarks related to policy formation in the future were added for a more comprehensive result. (Yazdani-Chamzini et al., 2013) proposed an AHP-COPRAS MCDM model for energy selection. The proposal also contained a list of results of comparison of their methods, notably TOPSIS, VIKOR, SAW, ARAS, and their corresponding effectiveness. (Sánchez-Lozano, et al., 2013) combined MCDM with the GIS optical locator to determine the best solar plant setup in Spain. (Bagočius et al., 2014) selected various viable locations for natural gas generators under the Baltic Sea. Subsequently, near by Islands were crowned as the most suitable location by SAW and TOPSIS. (Chen et al., 2014) determined the most important criteria for the construction of solar power plants in China. Average temperature, distance from habitat, and solar radiation intensity were classified as the top three critical criteria. (Georgiou et al., 2015) implemented the combination of AHP and PROMETHEE to find the most effective desalination technique via reverse osmosis. Furthermore, the direct-coupled system was selected as the supreme alternative, considering a slew of issues like environmental factors, economic burden, and ecosystem impact due to the constructions. (Ren et al., 2009) compared low carbon emission energy systems using AHP and TOPSIS. Nuclear and solar were rated as less desirable as compared to wind and hydroelectric sources. (Ghosh et al., 2016) approached the problem from a machine learning perspective: Using an artificial neural network to locate sources of sinusoidal energy generation. (Büyüközkan & Güleryüz, 2016) selected wind energy as the most beneficial non-renewable source of energy for investors based in Turkey, with a DEMATEL-based evaluation system.

5.4 SCRUTINY ANALYSIS OF PAPERS ABOUT ENERGY MANAGEMENT IN POLICY-MAKING

The assessment and analysis of policy-oriented decisions and assaying current or previously applied strategies is an abstruse task that can be handled by exploring various dimensions according to the deliberated attributes and variegated aims or objectives. This scenario can be considered suitable for the application of Multi-Criteria Decision-Making (MCDM) strategies by exploring various attributes, such as technological advancements, available resources, and diverse objectives along with disparate criteria. Hence, for solving, analyzing, and evaluating decision-making policy related to energy, MCDM methods and algorithms have proven great efficacy in the current, available literature. The proposed work and literature provide an insightful view of multiple MCDM methods, their respective applications, and advancements from the perspective of decision-making in the field of energy with recommendations for future scope. Multi-criteria decision-making methodologies that aid in assaying various alternatives based on multiple custom attributes can be graded easily.

These methods can be broadly described as comparison-based approaches, i.e. AHP and ANP, methods based on distance approaches, i.e. TOPSIS and VIKOR, while outranking approaches are ELECTRE, PROMETHEE, and DEMATEL. Miscellaneous approaches include Choquet Integral. For comprehending weight calculations for criteria, two popularly used methods are the Analytic Hierarchy Process (AHP) and the Analytic Network Process (ANP), which have various advantages on the grounds of monetary funds, risk evaluation, and scope. The disadvantage of the two methods is mainly in calculating the consistency ratio that quantizes or validates if the comparison performed is authentic, but these methods can also be implemented when an alternative analysis is done, as its efficacy is proven in this area. On the contrary, the methods using grade available alternatives based on their proximity for the best ideal case solution (by utilizing any germane method) are the Technique for Order of Preference by Similarity to Ideal Solution (TOPSIS) and Vlse Kriterijumska Optimizacija I Kompromisno Resenje (VIKOR). Some MCDM methodologies are implemented with the aim of developing an alternatives triage, i.e. prioritizing them by their intermittent ranking relations, their primordial aim to draw a comparison between alternatives and create a prioritized list. Approaches like the decision-making trial and evaluation laboratory (DEMATEL) are specifically used to analyze and rank intermittent relationships among the deliberated attributes to winnow the attributes selected for application of further analysis. Deliberating various methods of MCDM, one cannot declare which method is universally preferable, as each has its merits and demerits. One method can be preferred over another considering a specific scenario and deliberated attributes or model used.

To understand the suitability of any method, one needs to assay generic concepts, like validity, suitability, purview, user experience, and specific factors like the number of attributes, criteria to deliberate, and content-specific issues. Conclusively, various Multi-Criteria Decision-Making (MCDM) approaches are utilized with great efficacy and fecundity, showing optimal results especially in the area of energy conservation and energy sustainability policies in India to provide tenable and pragmatic solutions to various energy-related issues, like energy plant development or energy conservation policies. This research study's objective is to provide a comprehensive and detailed view of literature and work done various by erudite sources, i.e. implementing MCDM techniques and tool set-in decision-making quandary problematic situations. MCDM techniques may be conventional and well researched in nature, but they can be surprisingly used to solve new and cutting-edge decision-making problems as well. Hence, a detailed and comprehensive analysis of the available current literature can be of great benefit for the subject at hand. This research study hereby presents a literature review assaying the various publications, studies, and research work collated according to the method of

MCDM utilized, the year the work was published, and the location of publication, along with publishing authorities' information. The output is formulated in a tabular format of lucid understanding and clear representation. After analyzing the presented literature review in the tabular form, one can clearly say that the most popular and widely used technique is the Analytic Hierarchy Process (AHP), which is incidentally implemented individually or amalgamated with some sister MCDM approach, sometimes even combined with two or more methods creating a hybrid AHP or hybrid advanced AHP methodology which is them implemented on variegates, yet uniquely on energy conservation or energy policy decision-making scenarios. Combing multiple MCDM techniques is rather mundane in solving disparate and challenging decision-making problems in the field of energy. This is done to cater to the melange of MCDM decisions that arises, since each specific case deliberated is different and abstruse. Multiple methodologies are suitable in order to obtain optimal results.

In the current section, the presented research has assayed various problems and difficulties occurring in the renewable energy field and has attempted to reevaluate these from a Multi-Criteria Decision-Making or MCDM perspective. This further appraises their attributes, characteristics, and other priority-based techniques to evaluate the currently published and presented work and literature in order to instigate and whet new perspectives. The assayed literature and deliberated literature include variegated types and sources, including conference papers, authorized journals, and calibrated articles, all belonging to the concept of renewable energy and related policies. The results are presented in Table 5.1 below.

5.5 CONCLUSION

This study has evaluated various existing literature and solutions with the MCDM perspective to provide probable solutions. More than 100 research materials, including papers, journals, *etc.*, are evaluated.

The research proposes a statistical model to evaluate, estimate, and identify various literature which discusses the topic of energy policy and renewable energy implemented by the application of the Multi-Criteria Decision-Making or MCDM method. This model integrates the concept of Multi-Criteria Decision-Making or MCDM with regression models to obtain results which are then evaluated by using three accuracy parametres like Mean Absolute Error and Mean Absolute percentage error. Besides renewable energy and policy-related issues, this study also evaluates various neighbouring energy issues like sustainable energy and decision-related issues from a Multi-Criteria Decision-Making perspective, and the deliberated results are presented in graphical and tabular form.

The quantity of studies arranged by applying the MCDM strategy is introduced, and the dissemination rate of distribution is likewise given. The Scientific Hierarchy Process is the most regularly applied MCDM technique for vital decision-making and policy-making issues. The acquired outcomes show that AHP can be utilized as a solitary strategy or alongside another method to assess choice issues. Coincidentally, the strategies TOPSIS and ANP follow this strategy separately.

TABLE 5.1
A Classification of Literature Analysis with Respect to the MCDM Methods and Energy Types

	Renewable	Non-Renewable	AHP	ANP	DMTL	ELCTR	PRM	TPSS	VKR	Others
Abudeif et al. 2015		√	√							
Afsordegan et al. 2014	√			√						√
Akash et al. 1999			√							
Akbari et al. 2017	√		√							
Al Garni et al. 2016	√		√							
Alidrisi and al-Sasi 2017								√		
Al-Yahyai et al. 2013	√		√							
Amer and Daim 2011	√		√							
Aras et al. 2004	√		√							
Balin et al. 2012										
Blanco et al al. 2017	√		√							
Bagočius et al. 2014	√	√							√	
Bojesen et al. 2015	√		√							
Boran et al. 2013			√	√						
Brand and Missaoui 2014	√							√		
Buyukozkan and Guleryuz 2016	√			√	√					
Catalina et al. 2011										
Cavallaro 2009	√						√			
Çelikbilek and Tuysuz 2016	√			√	√				√	
Chandrasekhar et al. 2013	√		√							√
Chatzimouratidis and Pilavachi 2012			√							

(Continued)

TABLE 5.1 (Continued)
A Classification of Literature Analysis with Respect to the MCDM Methods and Energy Types

	Renewable	Non-Renewable	AHP	ANP	DMTL	ELCTR	PRM	TPSS	VKR	Others
Chatzimouratidis and Pilavachi 2012	√									
Chen et al. 2014								√		√
Daim et al. 2012			√							
Daim et al. 2012			√					√		√
Demirtas 2013	√		√							
Doukas et al. 2016							√			
Erdoğmuş et al. 2006		√		√						
Erol and Kılkıs 2012	√		√							√
Georgiou et al. 2015	√		√				√			
Georgopoulou et al. 1997	√					√				
Georgopoulou et al. 1998						√				
Georgopoulou et al. 2003	√						√			
Ghafghazi et al. 2010	√									
Ghosh et al. 2016	√		√							√
Goletsis et al. 2003						√	√			
Haddad et al. 2017	√		√							
Haralambopoulos and Polatidis 2003	√						√			
Kabak and Dagdeviren 2014	√						√			
Kablan 2004			√							
Kagazyo et al. 1997			√							
Karakosta et al. 2009						√				
Kim et al. 1999		√	√							

(Continued)

TABLE 5.1 (Continued)
A Classification of Literature Analysis with Respect to the MCDM Methods and Energy Types

	Renewable	Non-Renewable	AHP	ANP	DMTL	ELCTR	PRM	TPSS	VKR	Others
Kolios et al. 2016	√							√		√
Kone and Buke 2007				√						
Kowalski et al. 2009	√						√			
Kylili et al. 2016							√			
Lee et al. 2008	√		√							
Lee et al. 2008	√		√							√
Lee et al. 1999			√							√
Lerche et al. 2019	√						√			
Madlener and Stagl 2005	√						√			
Madlener et al. 2009	√						√			
Malkawi et al. 2017			√							
Önüt et al. 2008		√		√						
Özcan et al. 2017	√		√					√		
Özkale et al. 2017	√						√			
Papadapoulos and Karagiannidis 2008	√					√				
Phdungsilp 2010			√	√						
Pilavachi et al. 2012			√							
Polatidis and Haralambopoulos 2007	√						√			
Ren and Sovacool 2009			√					√		
Ren et al. 2015			√				√			
Rosso et al. 2014	√		√							√
Roy and Bouyssou 1986	√					√				

(Continued)

TABLE 5.1 (Continued)
A Classification of Literature Analysis with Respect to the MCDM Methods and Energy Types

	Renewable	Non-Renewable	AHP	ANP	DMTL	ELCTR	PRM	TPSS	VKR	Others
Sagbansua and Balo 2017	√		√							
Sánchez--Lozano et al. 2013			√							
Sánchez-Lozano et al. 2013	√		√			√		√		
Shen et al. 2011	√		√						√	
Shirgholami et al. 2016	√		√							
Shiue and Lin 2012							√			
Sindhu et al. 2017	√		√							
Streimikiene et al. 2018	√			√						√
Supriyasilp et al. 2009	√		√							
Tahri et al. 2015	√		√							
Talaei et al. 2014	√		√					√		
Terrados et al. 2009	√						√			
Toossi et al. 2013			√					√		
Topcu and Ulengin 2004							√			
Troldborg et al. 2014	√			√	√					√
Tsoutsos et al. 2009	√						√			
Tzeng et al. 2005			√					√	√	
Ulutas 2005				√						
Wang et al. 2010			√							
Wang et al. 2008			√				√			
Xiaohua and Zhenmin 2002	√		√							
Yazdani-Chamzini et al. 2013	√		√					√		
Zhao et al. 2009			√							

The various assayed energy alternatives among renewable and non-renewable energy include power plants quantifying and taping multiple energy sources like nuclear, solar, and wind, location selection for plant establishment (which itself is a cumbersome task considering and evaluating various criteria like geographical location), weather condition, labour, and other pivotal conditions; deliberating other energy evaluation criteria and their variegated sources, like various irreplaceable or non-negotiable conditions that are unique to every individual energy plants and its establishments, maintenance, and future development. All of these alternatives and criteria have to be thoroughly assessed before, after, and also during the plant establishment. Thereby, the Multi-Criteria Decision-Making perspective is explored in order to solve and comprehend this. These sub-criteria are often quite disparate in nature and cannot be graded, so it has become really enigmatic to deliberate and consider these all, as they belong to various fields like technical issues, pecuniary criteria, socio-political problems, and often create endless issues which become difficult handle simultaneously. However, these criteria are often quite important to avoid; thus, the amalgamation of multiple Multi-Criteria Decision-Making approaches and algorithms are implemented to provide a viable solution.

In addition, the MCDM methods identified with vitality have expanded as of late, since vitality arranging is one of the most appealing issues, and it is normal that it will likewise increase in the next years because of the developing populace and modern exercises. In the past few decades, the concept of 'Energy Planning' has experienced great attention and advancement, so the application of MCDM techniques has also shown sublime work. Different developing and developed countries like the United Kingdom, India, Iran, China, Turkey, and the United States have exhibited advancements both in terms of research and implementation. In addition, articles, meeting papers, and parts of books are kinds of distribution identified with vitality MCDM issues. Articles are considered as the broadest archive type for vitality MCDM. The implementation of MCDM is not limited to energy conservation and energy policy decision-making problems, but is also expanding its reach to new avenues like sustainable energy reviews, renewable energy, energy, applied energy scenarios, problem statements, and decision-making situations. For future reference and in-depth analysis, the mentioned literature can be referred to. However, according to trends observed in the literature, the two fields that have present groundbreaking improvements are the field of renewable and sustainable energy reviews, as they have managed to publish the most in different publications, thus solving multiple problems and exploring multifarious avenues.

For future research and scope, the application, advancements, concepts, and purview of Fuzzy MCDM methods will be explored. Thereby, the future will touch upon the topic of finding the most suitable MCDM-fuzzy technique both on grounds of applications and research.

REFERENCES

Abudeif, A.M., Moneim, A.A., & Farrag, A.F. (2015). Multicriteria decision analysis based on analytic hierarchy processing GIS environment for siting nuclear power plant in Egypt. *Annals of Nuclear Energy*, *75*, 682–692.

Afsordegan, A., Sánchez, M., Agell, N., Aguado, J.C., & Gamboa, G. (2014, October). A comparison of two MCDM methodologies in the selection of a windfarm location in Catalonia. In Museros L., et al. (Eds). *Artificial Intelligence Research and Development* (pp. 227–236). IOS Press.

Akash, B.A., Mamlook, R., & Mohsen, M.S. (1999). Multi-criteria selection of electric power plants using analytical hierarchy process. *Electric Power Systems Research*, *52*(1), 29–35.

Akbari, N., Irawan, C.A., Jones, D.F., & Menachof, D. (2017). A multi-criteria port suitability assessment for developments in the offshore wind industry. *Renewable Energy*, *102*, 118–133.

Alidrisi, H., & Al-Sasi, B.O. (2017). Utilization of energy sources by G20 countries: A TOPSIS-BASED approach. *Energy Sources, Part B: Economics, Planning, and Policy*, *12*(11), 964–970.

Al Garni, H., Kassem, A., Awasthi, A., Komljenovic, D., & Al-Haddad, K. (2016). A multicriteria decision making approach for evaluating renewable power generation sources in Saudi Arabia. Sustainable energy technologies and assessments, *16*, 137–150.

Al-Yahyai, S., Charabi, Y., Al-Badi, A., & Gastli, A. (2013). Wind resource assessment using numerical weather prediction models and multi-criteria decision making technique: Case study (Masirah Island, Oman). *International Journal of Renewable Energy Technology*, *4*(1), 17–33.

Amer, M., & Daim, T.U. (2011). Selection of renewable energytech nologiesfora developing county: A case of Pakistan. *Energy for Sustainable Development*, *15*(4), 420–435.

Aras, H., Erdoğmuş, Ş., & Koç, E. (2004). Multi-criteria selection for a wind observation station location using analytic hierarchy process. *Renewable Energy*, *29*(8), 1383–1392.

Arce, M.E., Saavedra, Á., Míguez, J.L., & Granada, E. (2015). The use of grey-based methods in multi-criteria decision analysis for the evaluation of sustainable energy systems: A review. *Renewable and Sustainable Energy Reviews*, *47*, 924–932.

Bagočius, V., Kazimieras Zavadskas, E., & Turskis, Z. (2014). Selecting a location for a liquefied natural gas terminal in the Eastern Baltic Sea. *Transport*, *29*(1), 69–74.

Beccali, M., Cellura, M., & Mistretta, M. (2003). Decision-making in energy planning. Application of the Electre method at regional level for the diffusion of renewable energy technology. *Renewable Energy*, *28*(13), 2063–2087.

Balin, A., Pelin A., & Hüseyin B. (2012). The applications of energy alternatives in Turkey using multicriteria decision making processes. In Uncertainty Modeling in Knowledge Engineering and Decision Making, pp. 124–130. https://doi.org/10.1142/9789814417

Behzadian, M., Kazemzadeh, R.B., Albadvi, A., & Aghdasi, M. (2010). PROMETHEE: A comprehensive literature review on methodologies and applications. *European Journal of Operational Research*, *200*(1), 198–215.

Blanco, G., Amarilla, R., Martinez, A., Llamosas, C., & Oxilia, V. (2017). Energy transitions and emerging economies: A multi-criteria analysis of policy options for hydropower surplus utilization in Paraguay. *Energy Policy*, *108*, 312–321.

Boemi, S.N., Papadopoulos, A.M., Karagiannidis, A., & Kontogianni, S. (2010). Barriers on the propagation of renewable energy sources and sustainable solid waste management practices in Greece. *Waste Management & Research*, *28*(11), 967–976.

Bojesen, M., Boerboom, L., & Skov-Petersen, H. (2015). Towards a sustainable capacity expansion of the Danish biogas sector. *Land Use Policy*, *42*, 264–277.

Boran, F.E., Etöz, M., & Dizdar, E. (2013). Is nuclear power an optimal option for electricity generation in Turkey?. *Energy Sources, Part B: Economics, Planning, and Policy, 8*(4), 382–390.

Brand, B., & Missaoui, R. (2014). Multi-criteria analysis of electricity generation mix scenarios in Tunisia. *Renewable and Sustainable Energy Reviews, 39*, 251–261.

Büyüközkan, G., & Güleryüz, S. (2016). An integrated DEMATEL-ANP approach for renewable energy resources selection in Turkey. *International Journal of Production Economics, 182*, 435–448.

Catalina, T., Virgone, J., & Blanco, E. (2011). Multi-source energy systems analysis using a multi-criteria decision aid methodology. *Renewable Energy, 36*(8), 2245–2252.

Cavallaro, F. (2009). Multi-criteria decision aid to assess concentrated solar thermal technologies. *Renewable Energy, 34*(7), 1678–1685.

Çelikbilek, Y., & Tüysüz, F. (2016). An integrated grey based multi-criteria decision making approach for the evaluation of renewable energy sources. *Energy, 115*, 1246–1258.

Chatzimouratidis, A.I., & Pilavachi, P.A. (2012). Decision support systems for power plants impact on the living standard. *Energy Conversion and Management, 64*, 182–198.

Chandrasekhar, V., Marthuvanan, M., Ramkumar, M. M., Shriram, R., Manickavasagam, V. M., Ramnath, B. V. (2013). MCDM Approach for Selecting Suitable Solar Tracking System. (pp. 148–152). IEEE. In 2013 7th International Conference on Intelligent Systems and Control.

Chen, C.R., Huang, C.C., & Tsuei, H.J. (2014). A hybrid MCDM model for improving GIS-based solar farms site selection. *International Journal of Photoenergy, 2014*, 1–9, https://doi.org/10.1155/2014/925370

Cheng, E. W., & Li, H. (2001). Analytic hierarchy process. *Measuring Business Excellence, 5*(3), 30–37.

Daim, T. U., Kayakutlu, G., & Cowan, K. (2010). Developing Oregon's renewable energy portfolio using fuzzy goal programming model. *Computers & Industrial Engineering, 59*(4), 786–793.

Daim, T. U., Li, X., Kim, J., & Simms, S. (2012). Evaluation of energy storage technologies for integration with renewable electricity: Quantifying expert opinions. *Environmental Innovation and Societal Transitions*, 3, 29–49.

Demirtas, O. (2013). Evaluating the best renewable energy technology for sustainable energy planning. *International Journal of Energy Economics and Policy, 3*, 23.

Doukas, H., Patlitzianas, K.D., & Psarras, J. (2006). Supporting sustainable electricity technologies in Greece using MCDM. *Resources Policy, 31*(2), 129–136.

Erdoğmuş, Ş., Aras, H., & Koç, E. (2006). Evaluation of alternative fuels for residential heating in Turkey using analytic network process (ANP) with group decision-making. *Renewable and Sustainable Energy Reviews, 10*(3), 269–279.

Erol, Ö., & Kılkış, B. (2012). An energy source policy assessment using analytical hierarchy process. *Energy Conversion and Management, 63*, 245–252.

Georgiou, D., Mohammed, E.S., & Rozakis, S. (2015). Multi-criteria decision making on the energy supply configuration of autonomous desalination units. *Renewable Energy, 75*, 459–467.

Georgopoulou, E., Lalas, D., & Papagiannakis, L. (1997). A multicriteria decision aid approach for energy planning problems: The case of renewable energy option. *European Journal of Operational Research, 103*(1), 38–54.

Georgopoulou, E., Sarafidis, Y., & Diakoulaki, D. (1998). Design and implementation of a group DSS for sustaining renewable energies exploitation. *European Journal of Operational Research, 109*(2), 483–500.

Georgopoulou, E., Sarafidis, Y., Mirasgedis, S., Zaimi, S., & Lalas, D.P. (2003). A multiple criteria decision-aid approach in defining national priorities for greenhouse gases emissions reduction in the energy sector. *European Journal of Operational Research, 146*(1), 199–215.

Ghafghazi, S., Sowlati, T., Sokhansanj, S., & Melin, S. (2010). A multicriteria approach to evaluate district heating system options. *Applied energy, 87*(4), 1134–1140.

Ghosh, S., Chakraborty, T., Saha, S., Majumder, M., & Pal, M. (2016). Development of the location suitability index for wave energy production by ANN and MCDM techniques. *Renewable and Sustainable Energy Reviews, 59*, 1017–1028.

Goletsis, Y., Psarras, J., & Samouilidis, J. E. (2003). Project ranking in the Armenian energy sector using a multicriteria method for groups. *Annals of Operations Research, 120*(1–4), 135–157.

Haddad, B., Liazid, A., & Ferreira, P. (2017). A multi-criteria approach to rank renewables for the Algerian electricity system. *Renewable Energy, 107*, 462–472.

Haralambopoulos, D. A., & Polatidis, H. (2003). Renewable energy projects: Structuring a multi-criteria group decision-making framework. *Renewable Energy, 28*(6), 961–973.

Kabak, M., & Dağdeviren, M. (2014). Prioritization of renewable energy sources for Turkey by using a hybrid MCDM methodology. *Energy Conversion and Management, 79*, 25–33.

Kablan, M.M. (2004). Decision support for energy conservation promotion: An analytic hierarchy process approach. *Energy Policy, 32*(10), 1151–1158.

Kagazyo, T., Kaneko, K., Akai, M., & Hijikata, K. (1997). Methodology and evaluation of priorities for energy and environmental research projects. *Energy, 22*(2–3), 121–129.

Karakosta, C., Doukas, H., & Psarras, J. (2009). Directing clean development mechanism towards developing countries' sustainable development priorities. *Energy for Sustainable Development, 13*(2), 77–84.

Kilic, M., & Kaya, İ. (2015). Investment project evaluation by a decision making methodology based on type-2 fuzzy sets. *Applied Soft Computing, 27*, 399–410.

Kim, P.O., Lee, K.J., & Lee, B.W. (1999). Selection of an optimal nuclear fuel cycle scenario by goal programming and the analytic hierarchy process. *Annals of Nuclear Energy, 26*(5), 449–460.

Kolios, A., Read, G., & Ioannou, A. (2016). Application of multi-criteria decision-making to risk prioritisation in tidal energy developments. *International Journal of Sustainable Energy, 35*(1), 59–74.

Köne, A.Ç., & Büke, T. (2007). An analytical network process (ANP) evaluation of alternative fuels for electricity generation in Turkey. *Energy Policy, 35*(10), 5220–5228.

Kowalski, K., Stagl, S., Madlener, R., & Omann, I. (2009). Sustainable energy futures: Methodological challenges in combining scenarios and participatory multi-criteria analysis. *European Journal of Operational Research, 197*(3), 1063–1074.

Kumar, A., Sah, B., Singh, A.R., Deng, Y., He, X., Kumar, P., & Bansal, R.C. (2017). A review of multi criteria decision making (MCDM) towards sustainable renewable energy development. *Renewable and Sustainable Energy Reviews, 69*, 596–609.

Kumar, K., Dhillon, V. S., Singh, P. L., & Sindhwani, R. (2019). Modeling and analysis for barriers in healthcare services by ISM and MICMAC analysis. In M. Kumar, R. Pandey, R. Kumar, (Eds). *Advances in Interdisciplinary Engineering*. Lecture Notes in Mechanical Engineering (pp. 501–510). Springer, Singapore. https://doi.org/10.1007/978-981-13-6577-5_47

Kumar, R., Kumar, V., and Singh, S. (2017a). Modeling and analysis on supply chain characteristics using ISM technique. *Apeejay Journal of Management and Technology, 12* (1 & 2), 21–30.

Kumar, R., Kumar, V., and Singh, S. (2017b). Work culture enablers: Hierarchical design for effectiveness & efficiency. *International Journal of Lean Enterprise Research (IJLER), 2*(3), 189–201.

Kylili, A., Christoforou, E., Fokaides, P.A., & Polycarpou, P. (2016). Multicriteria analysis for the selection of the most appropriate energy crops: The case of Cyprus. *International Journal of Sustainable Energy*, *35*(1), 47–58.

Lee, S.K., Mogi, G., Shin, S.C., & Kim, J.W. (2008, March). Measuring the relative efficiency of greenhouse gas technologies: An AHP/DEA hybrid model approach. In Proceedings of the (Vol. 2, pp. 1615–1619). International MultiConference of Engineers and Computer Scientists.

Lerche, N., Wilkens, I., Schmehl, M., Eigner-Thiel, S., & Geldermann, J. (2019). Using methods of multi-criteria decision making to provide decision support concerning local bioenergy projects. *Socio-Economic Planning Sciences*, *68*, 100594.

Madlener, R., & Stagl, S. (2005). Sustainability-guided promotion of renewable electricity generation. *Ecological Economics*, *53*(2), 147–167.

Malkawi, S., & Azizi, D. (2017). A multi-criteria optimization analysis for Jordan's energy mix. *Energy*, *127*, 680–696.

Mittal, V.K., Sindhwani, R., Shekhar, H., & Singh, P. L. (2019). Fuzzy AHP model for challenges to thermal power plant establishment in India. *International Journal of Operational Research*, *34*(4), 562–581.

Mittal, V.K., Sindhwani, R., Singh, P.L., Kalsariya, V., & Salroo, F. (2018). Evaluating significance of green manufacturing enablers using MOORA method for Indian manufacturing sector. In S. Singh, P. Raj, & S. Tambe (Eds.) Proceedings of the *International Conference on Modern Research in Aerospace Engineering* (pp. 303–314). Springer, Singapore.

Nigim, K., Munier, N., & Green, J. (2004). Pre-feasibility MCDM tools to aid communities in prioritizing local viable renewable energy sources. *Renewable Energy*, *29*(11), 1775–1791.

Önüt, S., Tuzkaya, U.R., & Saadet, N. (2008). Multiple criteria evaluation of current energy resources for Turkish manufacturing industry. *Energy Conversion and Management*, *49*(6), 1480–1492.

Özcan, E.C., Ünlüsoy, S., & Eren, T. (2017). A combined goal programming–AHP approach supported with TOPSIS for maintenance strategy selection in hydroelectric power plants. *Renewable and Sustainable Energy Reviews*, *78*, 1410–1423.

Özkale, C., Celik, C., Turkmen, A.C., & Cakmaz, E.S. (2017). Decision analysis application intended for selection of a power plant running on renewable energy sources. *Renewable and Sustainable Energy Reviews*, *70*, 1011–1021.

Papadopoulos, A., & Karagiannidis, A. (2008). Application of the multi-criteria analysis method Electre III for the optimisation of decentralised energy systems. *Omega*, *36*(5), 766–776.

Phdungsilp, A. (2010). Integrated energy and carbon modeling with a decision support system: Policy scenarios for low-carbon city development in Bangkok. *Energy Policy*, *38*(9), 4808–4817.

Pohekar, S.D., & Ramachandran, M. (2004). Application of multi-criteria decision making to sustainable energy planning—A review. *Renewable and Sustainable Energy Reviews*, *8*(4), 365–381.

Polatidis, H., & Haralambopoulos, D.A. (2007). Renewable energy systems: A societal and technological platform. *Renewable Energy*, *32*(2), 329–341.

Ren, H., Gao, W., Zhou, W., & Nakagami, K.I. (2009). Multi-criteria evaluation for the optimal adoption of distributed residential energy systems in Japan. *Energy Policy*, *37*(12), 5484–5493.

Rosso, M., Bottero, M., Pomarico, S., La Ferlita, S., & Comino, E. (2014). Integrating multicriteria evaluation and stakeholders analysis for assessing hydropower projects. *Energy Policy*, *67*, 870–881.

Roy, B., & Bouyssou, D. (1986). Comparison of two decision-aid models applied to a nuclear power plant siting example. *European Journal of Operational Research*, *25*(2), 200–215.

Sagbansua, L., & Balo, F. (2017). Decision making model development in increasing wind farm energy efficiency. *Renewable Energy*, *109*, 354–362.

Sánchez-Lozano, J.M., Teruel-Solano, J., Soto-Elvira, P.L., & García-Cascales, M.S. (2013). Geographical Information Systems (GIS) and Multi-Criteria Decision Making (MCDM) methods for the evaluation of solar farms locations: Case study in south-eastern Spain. *Renewable and Sustainable Energy Reviews*, *24*, 544–556.

Şengül, Ü., Eren, M., Shiraz, S.E., Gezder, V., & Şengül, A.B. (2015). Fuzzy TOPSIS method for ranking renewable energy supply systems in Turkey. *Renewable Energy*, *75*, 617–625.

Shanker, K., Shankar, R., & Sindhwani, R. (Eds.). (2019). *Advances in industrial and production engineering: Select proceedings of FLAME 2018*. Springer.

Shen, Y. C., Chou, C. J., Lin, G. T. (2011). The portfolio of renewable energy sources for achieving the three E policy goals. *Energy*, *36*(5), 2589–2598.

Shirgholami, Z., Zangeneh, S.N., & Bortolini, M. (2016). Decision system to support the practitioners in the wind farm design: A case study for Iran mainland. *Sustainable Energy Technologies and Assessments*, *16*, 1–10.

Shiue, Y.C., & Lin, C.Y. (2012). Applying analytic network process to evaluate the optimal recycling strategy in upstream of solar energy industry. *Energy and Buildings*, *54*, 266–277.

Sindhu, S., Nehra, V., & Luthra, S. (2017). Solar energy deployment for sustainable future of India: Hybrid SWOC-AHP analysis. *Renewable and Sustainable Energy Reviews*, *72*, 1138–1151.

Sindhwani, R. and Malhotra, V. (2017). A framework to enhance agile manufacturing system: A total interpretive structural modelling (TISM) approach. *Benchmarking: An International Journal*, *24*(2), 467–487. https://doi.org/10.1108/BIJ-09-2015-0092

Sindhwani, R., Mittal, V.K., Singh, P.L., Kalsariya, V., & Salroo, F. (2018). Modelling and analysis of energy efficiency drivers by fuzzy ISM and fuzzy MICMAC approach. *International Journal of Productivity and Quality Management*, *25*(2), 225–244.

Sindhwani, R., Singh, P.L., Chopra, R., Sharma, K., Basu, A., Prajapati, D.K., & Malhotra, V. (2019). Agility evaluation in the rolling industry: A case study. In K. Shankar, R. Shankar, & R. Sindhwani (Eds.). *Advances in industrial and production engineering*. Lecture Notes in Mechanical Engineering (pp. 753–770). Springer, Singapore.

Sindhwani, R., Singh, P.L., Iqbal, A., Prajapati, D.K., & Mittal, V.K. (2019). Modeling and analysis of factors influencing agility in healthcare organizations: An ISM approach. In K. Shankar, R. Shankar, & R. Sindhwani (Eds.). *Advances in industrial and production engineering*. Lecture Notes in Mechanical Engineering (pp. 683–696). Springer, Singapore.

Strantzali, E., & Aravossis, K. (2016). Decision making in renewable energy investments: A review. *Renewable and Sustainable Energy Reviews*, *55*, 885–898.

Streimikiene, D., Siksnelyte, I., Zavadskas, E.K., & Sharma, D. (2018). An overview of multi-criteria decision-making methods in dealing with sustainable energy development issues. *Energies*, *11*(10), 2754.

Supriyasilp, T., Pongput, K., & Boonyasirikul, T. (2009). Hydropower development priority using MCDM method. *Energy Policy*, *37*(5), 1866–1875.

Suganthi, L., Iniyan, S., & Samuel, A. A. (2015). Applications of fuzzy logic in renewable energy systems–A review. *Renewable and Sustainable Energy Reviews*, *48*, 585–607.

Tahri, M., Hakdaoui, M., & Maanan, M. (2015). The evaluation of solar farm location sapplying geographic information system and multi-criteria decision-making methods: Case study in southern Morocco. *Renewable and Sustainable Energy Reviews*, *51*, 1354–1362.

Talaei, A., Ahadi, M.S., & Maghsoudy, S. (2014). Climate friendly technology transfer in the energy sector: A case study of Iran. *Energy Policy*, *64*, 349–363.

Terrados, J., Almonacid, G., & PeRez-Higueras, P. (2009). Proposal for a combined methodology for renewable energy planning. Application to a Spanish region. *Renewable and Sustainable Energy Reviews*, *13*(8), 2022–2030.

Topcu, Y.I., & Ulengin, F. (2004). Energy for the future: An integrated decision aid for the case of Turkey. *Energy*, *29*(1), 137–154.

Toossi, A., Camci, F., & Varga, L. (2013, February). Developing an AHP based decision model for energy systems policymaking. (pp. 1456–1460). IEEE. In 2013 IEEE International Conference on Industrial Technology(ICIT)

Troldborg, M., Heslop, S., & Hough, R.L. (2014). Assessing the sustainability of renewable energy technologies using multi-criteria analysis: Suitability of approach for national-scale assessments and associated uncertainties. *Renewable and Sustainable Energy Reviews*, *39*, 1173–1184.

Tsoutsos, T., Drandaki, M., Frantzeskaki, N., Iosifidis, E., & Kiosses, I. (2009). Sustainable energy planning by using multi-criteria analysis application in the island of Crete. *Energy Policy*, *37*(5), 1587–1600.

Tzeng, G.H., Lin, C.W., & Opricovic, S. (2005). Multi-criteria analysis of alternative-fuel buses for public transportation. *Energy Policy*, *33*(11), 1373–1383.

Ulutaş, B.H. (2005). Determination of the appropriate energy policy for Turkey. *Energy*, *30*(7), 1146–1161.

Wang, B., Kocaoglu, D.F., Daim, T.U., & Yang, J. (2010). A decision model for energy resource selection in China. *Energy Policy*, *38*(11), 7130–7141.

Wang, J.J., Jing, Y.Y., Zhang, C.F., Zhang, X.T., & Shi, G.H. (2008). Integrated evaluation of distributed triple-generation systems using improved grey incidence approach. *Energy*, *33*(9), 1427–1437.

Wibowo, S., & Grandhi, S. (2019). A multicriteria group decision making approach for evaluating renewable power generation sources. In. R. Lee (Ed.). *Computer and Information Science. ICIS 2018. Studies in Computational Intelligence* (pp. 75–86). Springer, Cham.

Xiaohua, W., & Zhenmin, F. (2002). Sustainable development of rural energy and its appraising system in China. *Renewable and Sustainable Energy Reviews*, *6*(4), 395–404.

Yazdani-Chamzini, A., Fouladgar, M.M., Zavadskas, E.K., & Moini, S.H.H. (2013). Selecting the optimal renewable energy using multi criteria decision making. *Journal of Business Economics and Management*, *14*(5), 957–978.

6 Assessing Challenges in the Implementation of Sustainable Human Resource Management Using the Fuzzy DEMATEL Approach

Vernika Agarwal, Snigdha Malhotra, and Tilottama Singh

CONTENTS

6.1 Introduction 107
6.2 Literature Review 109
6.2.1 Sustainable HRM 109
6.2.2 Sustainable Supply Chain Management 109
6.2.3 Integration of HRM Practices and SCM 109
6.2.4 Challenges and Issues in Implementing Sustainability 110
6.3 Methodology 111
6.3.1 Fuzzy DEMATEL Approach 111
6.3.2 Problem Environment 113
6.3.3 Numerical Illustration 114
6.4 Discussion and Conclusion 114
References 118

6.1 INTRODUCTION

The international business world has been evolving continuously to overcome the challenges to its survival. The increasing pressure of stakeholders towards the inclusion of social sustainability along with the intense competition in the market is forcing businesses to understand the impact of sustainability for long-term success. A large number of businesses have started adopting the 'the triple bottom line' of sustainability into their operations, which aims to examine the company's social, environmental, and economic impact. As per the report by (Whelan &

Fink, 2016), it was stated that, at present, the value of spent by companies from sustainability concerns can be as high as 70% of earnings before interest, taxes, depreciation, and amortization. This ever-changing business world has confronted new challenges in business operations by simultaneously incorporating economic, ecological, and social values (Slaper & Hall, 2011). In the recent past, the terms sustainability,' 'sustainable development', 'corporate sustainability,' and 'corporate social responsibility' have been used interchangeably.

When talking about sustainability, one cannot oversee the function of human assets in any organisation which aims to build and preserve competitive advantage (Sindhwani et al., 2019; Wirtenberg et al., 2007). The human resource (HR) department incorporates all of the steps in the supply chain necessary in meeting the social sustainability objective. For HR practitioners, there are five key areas incorporating sustainability measures, which include planning, talent management and strategic human resource management, leadership, and improvement of workers (Wirtenberg et al., 2007). According to the literature, there is an alarming dearth in the area of 'Sustainable Human Resource Management' (Ehnert et al., 2014), and there is inadequate research done in this field as observed in its embryonic state. Hence, according to (De Prins et al., 2014), research in the area of Sustainable HRM is in the emerging phase, and concepts on how sustainability can be used constructively for HRM, as well as methods for use in practice, are being generated. Several terms have been used to link HRM with sustainability, such as sustainable work systems (Docherty et al., 2002), human resource sustainability (Wirtenberg et al., 2007), and sustainable management of human resources (Ehnert et al., 2014).

Once the current literature on sustainable HR practices in the manufacturing industry supply chain has been evaluated, this study attempts to bridge this research gap by addressing specific issues in the successful implementation of various methods in the construction of sustainable supply chain dynamics in the manufacturing industry. The study aims in understanding the major challenges faced by HRM, such as prior studies, but it also uses the multi-criteria decision-making method of fuzzy decision-making trial and evaluation laboratory (DEMATEL) to determine the interrelationship among these challenges and to identify them into a cause-and-effect group. In recent years, several previous studies have employed MCDM tools and applications to solve area problems such as energy, environment, and sustainability. The results of the research might be beneficial for growing manufacturing industries that are aiming for the successful establishment of sustainability driving the profits.

The initial section of the paper explains the concepts of Sustainable Human Resources in the manufacturing industry and relationships among supply chain, HRM, and operations of the business. The subsequent part of the paper explores the literature related to sustainable supply chain human resource management and supports a new approach with its rationality in the business. This is trailed by a conclusion and further recommended developments in research. This study is devoted to the specific sustainable practices in various dimensions by using the MCDM approach of DEMATEL which helps in identifying the interrelationships between these challenges and classifying them into cause-and-effect groups.

6.2 LITERATURE REVIEW

The literature was reviewed to gain an understanding of the challenges faced by organisations in aligning Human Resource practices for successfully implementing sustainable supply chains in the company. In this view, the literature review has been divided into sections explaining Sustainable Human Resource Management: sustainable supply chain management, integration of SHRM and SSCM, and the issues involved in implementing sustainability.

6.2.1 Sustainable HRM

Since the organisations are acknowledging the increasing importance of sustainability, the attention has widely shifted towards its implementation in business scenarios. Sustainable HRM helps to develop suitable conditions for individual employee sustainability and develops the ability of HRM systems to function effectively. According to Ehnert et al. (2014), by making the HRM system sustainable, companies can continuously attract and develop a motivated and engaged workforce. It also aligns with top management and helps organisations realize economic, human, social, and ecological sustainability goals, which ensures the sustainability of the business organisations in all facets. Research has shown that Sustainable HRM may wield significant benefits during global uncertainty and helps organisations to nullify the impact of policies like cost-cutting, decrease in capacity building, *etc* during times of crisis. (Maley & Kramar, 2015)

6.2.2 Sustainable Supply Chain Management

Business enterprises are increasingly focusing on buying and supplying products and services that reduce the adverse effects on the environment, society, and economy. By managing and improving environmental, social, and economic performance throughout supply chains, companies can conserve resources, optimize processes, uncover product innovations, save costs, increase productivity, and promote corporate values (Kumar et al., 2017a; Mittal et al., 2018; Sindhwani et al., 2018). Research shows that in the case of business, supply chain sustainability is growing. Business firms are under tremendous pressure to able to sustain their existing SC due to recent trends of globalization, market changes, demand uncertainty, and economic challenges. Focusing only stressing on the internal efficiencies of SC will be insufficient to gain a competitive advantage. If sustainability concepts are integrated into the core functions of a business firm's SC, then it achieves a good international market position (Khodakarami et al., 2015).

6.2.3 Integration of HRM Practices and SCM

Researchers are highlighting the exploration of the roles of SCM and HRM in creating more sustainable organisations. (Teixeira et al., 2016) proposed an integrative framework for the addition of contemporary fields to help organisations produce green and sustainable results. (Kumar et al., 2019) considered an example of an automotive company in India in his research to construct a structural framework for identifying the importance of the soft dimensions in adopting Green Supply Chain Management. The result shows that, for efficient Green SCM, the highly prioritized dimensions are 'teamwork,' 'employee involvement,' 'top management commitment,' and 'organisational culture.' (Florescu et al., 2019), in her study on the oil and gas distribution industry, analyzed the effect of SSCM strategies like Logistics Management,' 'Supplier Selection,' and 'Product Stewardship' on SCM functions like Planning, Execution, Coordination, and Collaboration. Her findings show that SSCM strategies have a positive and significant influence on Supply Chain Management functions. (Das, 2018) investigated that SSCM practices have an association with the environmental, social, and operational performance of the enterprise. A study has been conducted including environmental management practices, socially inclusive practices for employees, socially inclusive practices for the community, operations practices, and supply chain integration as exogenous variables. Organisational performance considered in the study is comprised of five dimensions: Environmental performance, employee-centred social performance, community-centred social performance, operations performance, and competitiveness as endogenous variables.

Ab Talib and Hamid (2014) stated that collaborative partnership, information technology, top management support, and human resource are critical success factors in supply chain management. Whereas, Al-Odeh and Smallwood (2012) stated that the effects of Sustainable SCM strategies are still unclear, and these may cause any type of outcome, positive or negative, on financial performance. His research highlights the advantages and barriers of SSCM. Anastasiou (2012) tried to derive important parametres for an effective Human Resources Management policy and potential practical steps for improvement in logistics as well as SCM. Because of this, companies display innovation, efficiency, flexible structure, and require new skills in their organisations and human resources.

6.2.4 Challenges and Issues in Implementing Sustainability

Manufacturing industries started adopting the green concept in their supply chain management to focus on environmental issues. In the industrial perspective, the industries are still struggling in identifying the barriers hindering green supply chain management implementation. Govindan et al., 2014 reported one case study on identified barriers and ranked the barriers based on MCDM approach. This took into account manufacturing companies and examined the relationship among different issues, like waste management, scarcity of raw materials, liquid waste discharge, groundwater level, environmentally friendly transportation system, production process, gaseous emissions, core manufacturing activities of supply chain, and the impact of product usage on the environment (Kumar et al., 2017b; Sriyogi et al., 2017).

As per the literature, a collective and comparative study to find the issues and challenges faced by HRM in achieving sustainability through establishing sustainable supply chains has not been conducted. With the motivation to cover the aforementioned gaps in the literature, the present research aims to identify the barriers faced by HRM in implementing sustainability in business operations and to prioritize these using the MCDM technique. Thus, two important contributions to the study are:

a. Identification of critical issues responsible as barriers for Human Resource Management for smooth the implementation of sustainable supply chains. The identification of barriers will support in creating a sustainable environment in developing nations.
b. The identified barriers are categorized and prioritized using a multi-criteria decision model, the Fuzzy DEMATEL method. The results have been evaluated and various policy recommendations to eliminate the significant barriers have been discussed.

6.3 METHODOLOGY

This section elaborates on the research methodology. In the present study, we have used the multi-criteria decision approach of fuzzy DEMATEL to understand the interrelationships amidst the factors of sustainable HRM. This MCDM approach offers cause-effect relationships to rank the attributes. The fuzzy set theory has been utilized to incorporate vagueness in the decision-making process (Gaur et al., 2020).

6.3.1 Fuzzy DEMATEL Approach

There is growing acceptance of SSCM practices amongst manufacturing and process-based organisations in India; thus, the firms need to align their HR practices with the supply chain in order to accelerate the involvement of members of the supply chain and ensure better business outcomes. Accordingly, the focus of the current study is to list the HRM issues and categorize these based on their contribution to sustainable supply chain management, providing a framework to the companies. These metrics are identified through literature review and detailed discussions with the DMs. To segment the factors identified through literature and categorize these into a cause-and-effect group, we have employed the DEMATEL methodology of multi-criteria decision analysis. To analyze the linguistic assessment by DMs, the authors have combined fuzzy set theory with DEMATEL. The steps of the integrated fuzzy DEMATEL approach are as follows (Agarwal et al., 2016; Vimal et al., 2020):

Step–1: Recognizing the decision body and short-listing factors. A decision-making body is formed to contribute to the identification of the goal of the study followed by short-listing the factors.

Step–2: Detecting the fuzzy linguistic scale. In this step, the scale for the evaluation of the metrics and the corresponding fuzzy numbers are identified.

Step–3: Determining the assessments of the decision-making body. The relationship among shortlisted metrics was determined to generate an initial direct matrix (IDM) in this step.

Step–4: Defuzzification of the IDM matrix. The linguistic assessments by the DMs were defuzzified into crisp value by using CFCS (i.e. Converting Fuzzy data into Crisp Scores) defuzzification (Opricovic & Tzeng, 2003). Let $Z_{ij} = (a_{ij}^l, b_{ij}^l, c_{ij}^l)$ be the effect of metric i on metric j for the lth DM. $Z_{ij} = (a_{ij}^l, b_{ij}^l, c_{ij}^l)$ represents a triangular fuzzy number (TFN) where $a_{ij}^l, b_{ij}^l, c_{ij}^l$ are left, middle, and right values. The steps of the CFCS method are as follows:

Step–4.1: *Normalisation*:

$$xa_{ij}^l = \frac{(a_{ij}^l - \min a_{ij}^l)}{\max c_{ij}^l - \min a_{ij}^l} \tag{6.1}$$

$$xb_{ij}^l = \frac{(b_{ij}^l - \min a_{ij}^l)}{\max c_{ij}^l - \min a_{ij}^l} \tag{6.2}$$

$$xc_{ij}^l = \frac{(c_{ij}^l - \min a_{ij}^l)}{\max c_{ij}^l - \min a_{ij}^l} \tag{6.3}$$

Step–4.2: *Determine the left (lxa) and right (lxc) normalized value:*

$$lxa_{ij}^l = \frac{xb_{ij}^l}{(1 + xb_{ij}^l - xa_{ij}^l)} \tag{6.4}$$

$$lxc_{ij}^l = \frac{xc_{ij}^l}{(1 + xc_{ij}^l - xb_{ij}^l)} \tag{6.5}$$

Step–4.3: *Calculate total normalized crisp value:*

$$x_{ij}^l = \frac{[lxa_{ij}^l(1 - lxa_{ij}^l) + lxc_{ij}^l lxc_{ij}^l]}{(1 - lxa_{ij}^l + lxc_{ij}^l)} \tag{6.6}$$

Step–4.4: *Calculate crisp value:*

$$k_{ij}^l = \min a_{ij}^l + x_{ij}^l(\max c_{ij}^l - \min a_{ij}^l) \quad (6.7)$$

Step–4.5: Determine the integrated value:

$$k_{ij} = \frac{1}{l}(k_{ij}^1 + k_{ij}^2 + \ldots + k_{ij}^l) \quad (6.8)$$

Step–5: Determining the normalized direct relation matrix. The DRM is normalized as:

$$K = [k_{ij}]_{nxn} = \frac{A}{s} \quad where \quad s = \max\left(\max_{1\le i\le n} \sum_{j=1}^{n} k_{ij}, \max_{1\le j\le n} \sum_{i=1}^{n} k_{ij}\right) \quad (6.9)$$

Step–6: Calculate the total relation matrix T:

$$T = [t_{ij}]_{nxn} = \lim_{m\to\infty} (K + K^2 + K^3 \ldots\ldots + K^m) = K(I - K)^{-1} \quad (6.10)$$

Step–7: Building the Casual Diagram

Let Ri and Cj be n × 1 and 1 × n matrices showing the row-sum and column-sum of T. Ri represents the total effects exerted by factor i on another factor, while Cj stands for the total effects received by factor j. Thus, (Ri + Ci) indicates the level of significance of the ith factor in the decision-making process. On the other hand, (Ri − Ci) indicates the resultant effect of the ith factor. To reduce the factors that bear a negligible effect on others, the DMs may generate a threshold limit. The factors with a value greater than the threshold limit will be chosen to map in the dataset (Ri + Ci, Ri − Ci) in the form of a causal diagram.

6.3.2 Problem Environment

Sustainable HRM is holistic and futuristic which aims at retention, regeneration, and renewal of professionals. A firm's human resource practices need to be aligned with its supply chain in order to foster the involvement of the members of the supply chain, to ensure the sustainability of the supply chain, and, consequently, to secure better business outcomes. Researchers and academicians around the globe have worked comprehensively in HRM. The growth of sustainability and the pressure of the stakeholders for the inclusion of sustainability into each aspect of the business process is forcing companies to become more sustainable. Although the focus of the Indian government and the stakeholders is towards sustainable development, its implementation is only at the budding stage. In this context, the paper aims to list the HRM issues and categorize these based on their contribution to sustainable supply chain management. The first step is to understand the challenges that are being faced by HRM in implementing sustainability in business operations. In this view, the current research attempts to comprehend sustainable HRM issues through the evidence in the literature and expert option. The identified challenges will be studied using the MCDM technique to understand their impact on one another. Based on the aforementioned study, policy measures and recommendations will also be proposed.

6.3.3 Numerical Illustration

A resource team that consists of representatives from manufacturing industries, non-government organisation volunteers, and environmental experts was formulated, where each member has 10–12 years of experience in the field. A workshop session was conducted with the resource team members. The nature and purpose of the study were explained to the members. Based on the input from the resource team members and through the literature study, 10 challenges for the implementation of sustainable HRM have been identified (Table 6.1). These challenges were accordingly confirmed with the resource team members. For interaction analysis, responses were collected from each member individually through face-to-face interviews to eliminate non-response. The collected responses were then processed and synthesized using the proposed method for obtaining the casual interrelationship model.

Using the steps of the fuzzy DEMATEL methodology, we will discover the relationships among sustainable HRM factors. The DMs were asked to provide justice for the pairwise comparison table based on a fuzzy linguistic scale. Table 6.2 provides the IDM assessment of the identified factors, as given by DMs. Once these initial relationships are identified by the consensus of all of the DMs, the IDM is then obtained using the CFCS method equations (6.1)–(6.8), as listed in Table 6.3.

After obtaining the IDM, the next step is to normalize it to generate the NDR matrix, as presented in Table 6.4. Following the next step of the research methodology, we obtain the total-relation matrix, as shown in Table 6.5. Next, we generate the causal diagram, as illustrated in Figure 6.1, by mapping a dataset of (Ci + Ri, Ci − Ri), as given in Table 6.6. On analyzing this causal diagram provided in Figure 6.1, it is clear that metrics were visually divided into the cause-and-effect group.

6.4 DISCUSSION AND CONCLUSION

The paper shows the causal dependency analysis of obstacles for the implementation of Sustainable Human Resource Management. Feedback from six manufacturers on their identification of barriers was gathered. After the review session with the authors, 10 barriers were finalized for the study. Multi-Criteria Decision Methods have been considered as an efficient system with excellent

TABLE 6.1
Challenges for the Implementation of Sustainable HRM

Notation	Challenges	Description	References
SH1	Health and safety incidents and practices	An effort to reduce the number of accidents on-site and taking precautions for the health and safety of employees. Having HR policies that ensure regular health check-ups.	Mokhtar et al. (2016)
SH2	Supporting community projects	Community involvement helps to bring positive measurable change to both the business and to the society in which the business exists and operates.	Mokhtar et al. (2016)
SH3	Work safety and labour health	Mechanism of workplace improvements at one site supported by regular training programs.	Mokhtar et al. (2016)
SH4	Percent of employment sourced from local communities	Reaching out for employment offers to disabled and lower-caste groups. Having a fixed percentage of employment of the identified categories.	Zaid et al. (2018)
SH5	Investment in community outreach	The case of corporate community involvement has never been stronger. Sharing a percentage of profit for community development.	Zaid et al. (2018)
SH6	Benefit shared with affected communities	Workers, natural environment, and communities improve by organisations practising social responsibility via contribution to sustainable development.	Hohenstein et al. (2014)
SH7	Human training and development	Training is an essential activity. Corporations need to invest in training of new, as well as old, employees on board.	Malhotra and Singh (2018)
SH8	Improvement in corporate image	First impressions are always important, and this is true for companies as well. The quality of products and services only matters if the image is correct.	Hohenstein, et al. (2014)
SH9	Funding donations and sponsorship activities	In-kind and financial donations, employee volunteer days, enduring non-profit partnerships, and other social activities are examples of community outreach practices adopted by companies.	Zaid et al. (2018)
SH10	Improvement in relations with community stakeholders (e.g. NGOs and community activists)	Active participation in non-profit partnerships, funding activities, and social volunteering helps organisations to build a positive corporate image.	Mokhtar et al. (2016)

TABLE 6.2
Linguistic Assessments by First Decision-Makers

	SH1	SH2	SH3	SH4	SH5	SH6	SH7	SH8	SH9	SH10
SH1	No	No	H	VL	VL	L	L	H	No	H
SH2	VL	No	No	H	VH	VH	VL	H	VL	VH
SH3	H	VL	No	VL	VL	VL	H	H	VL	VL
SH4	VL	H	L	No	VH	H	VL	H	L	H
SH5	VL	H	H	H	No	H	VL	L	VL	H
SH6	L	H	L	H	H	No	VL	H	L	H
SH7	H	L	H	VL	VL	L	No	H	H	L
SH8	H	H	H	H	H	H	L	No	L	L
SH9	VL	H	VL	VL	VL	H	L	H	No	H
SH10	L	H	VL	H	VH	H	VL	H	L	No

TABLE 6.3
Integrated Crisp Initial Decision Matrix

	SH1	SH2	SH3	SH4	SH5	SH6	SH7	SH8	SH9	SH10
SH1	0.3340	0.4254	0.8643	0.3917	0.3004	0.4254	0.6417	0.7667	0.3340	0.7667
SH2	0.5167	0.3340	0.3004	0.7667	0.8643	0.6143	0.2667	0.6417	0.3004	0.6480
SH3	0.7667	0.3004	0.3340	0.2667	0.3004	0.3004	0.6417	0.5167	0.3004	0.3004
SH4	0.2667	0.5504	0.3917	0.3340	0.6143	0.5167	0.3917	0.5167	0.4254	0.5167
SH5	0.3917	0.6417	0.5167	0.5167	0.3340	0.5167	0.3917	0.3917	0.2667	0.5167
SH6	0.3917	0.7667	0.5167	0.5167	0.6417	0.3340	0.3917	0.6417	0.3917	0.5504
SH7	0.5167	0.3917	0.6417	0.3004	0.3004	0.4254	0.3340	0.5167	0.5504	0.3917
SH8	0.6417	0.5167	0.6417	0.5504	0.5504	0.5504	0.3917	0.3340	0.4254	0.4254
SH9	0.3004	0.5167	0.3004	0.3004	0.3004	0.5167	0.3917	0.5167	0.3340	0.5167
SH10	0.3917	0.5167	0.2667	0.5167	0.6143	0.5167	0.3917	0.7667	0.4254	0.3340

potential for an effective decision-making process. Thus, this paper examines the use of the MCDM approach of fuzzy DEMATEL. The fuzzy DEMATEL method-based solution framework was used for the analysis. The interrelationships among the barriers were demonstrated through the results. Particularly, this method can also successfully divide the factors into cause-and-effect groups through a causal diagram; thus, the complexity of a problem becomes easy to interpret, and future decisions can be made. This paper provides critical criteria among various influencing elements. The results from the causal diagram divide criteria into two groups of cause and effect. SH1 – 'health and safety incidents and practices,' SH2 – 'supporting community projects,' SH4 – 'percent of employment sourced from local communities,' SH6 – 'benefit shared with affected communities,' SH7 – 'human training and

TABLE 6.4
The NDR Matrix

	SH1	SH2	SH3	SH4	SH5	SH6	SH7	SH8	SH9	SH10
SH1	0.0595	0.0758	0.1541	0.0698	0.0535	0.0758	0.1144	0.1367	0.0595	0.1367
SH2	0.0921	0.0595	0.0535	0.1367	0.1541	0.1095	0.0475	0.1144	0.0535	0.1155
SH3	0.1367	0.0535	0.0595	0.0475	0.0535	0.0535	0.1144	0.0921	0.0535	0.0535
SH4	0.0475	0.0981	0.0698	0.0595	0.1095	0.0921	0.0698	0.0921	0.0758	0.0921
SH5	0.0698	0.1144	0.0921	0.0921	0.0595	0.0921	0.0698	0.0698	0.0475	0.0921
SH6	0.0698	0.1367	0.0921	0.0921	0.1144	0.0595	0.0698	0.1144	0.0698	0.0981
SH7	0.0921	0.0698	0.1144	0.0535	0.0535	0.0758	0.0595	0.0921	0.0981	0.0698
SH8	0.1144	0.0921	0.1144	0.0981	0.0981	0.0981	0.0698	0.0595	0.0758	0.0758
SH9	0.0535	0.0921	0.0535	0.0535	0.0535	0.0921	0.0698	0.0921	0.0595	0.0921
SH10	0.0698	0.0921	0.0475	0.0921	0.1095	0.0921	0.0698	0.1367	0.0758	0.0595

TABLE 6.5
Total Relation Matrix

	SH1	SH2	SH3	SH4	SH5	SH6	SH7	SH8	SH9	SH10
SH1	0.5363	0.5813	0.6442	0.5333	0.5544	0.5592	0.5500	0.7087	0.4455	0.6379
SH2	0.5649	0.5824	0.5533	0.6116	0.6665	0.6035	0.4871	0.6945	0.4429	0.6317
SH3	0.5032	0.4467	0.4502	0.4070	0.4387	0.4296	0.4546	0.5388	0.3539	0.4501
SH4	0.4564	0.5428	0.4949	0.4677	0.5489	0.5153	0.4449	0.5892	0.4092	0.5331
SH5	0.4776	0.5547	0.5156	0.4976	0.4992	0.5125	0.4451	0.5680	0.3803	0.5320
SH6	0.5365	0.6397	0.5763	0.5571	0.6156	0.5427	0.4978	0.6803	0.4490	0.6010
SH7	0.4863	0.4924	0.5245	0.4397	0.4694	0.4789	0.4242	0.5701	0.4177	0.4927
SH8	0.5670	0.5824	0.5887	0.5458	0.5810	0.5644	0.4905	0.6135	0.4448	0.5668
SH9	0.4178	0.4873	0.4330	0.4171	0.4466	0.4687	0.4037	0.5364	0.3576	0.4849
SH10	0.4999	0.5616	0.4997	0.5219	0.5726	0.5390	0.4653	0.6584	0.4278	0.5264

development,' SH9 – 'funding donations and sponsorship activities (which belong to the cause group that should be controlled and given more attention),' SH3 – 'work safety and labour health,' SH5 – 'investment in community outreach,' SH8 – 'improvement in corporate image,' and SH10 – 'improvement in relations with community stakeholders, e.g. NGOs and community activists (which are in the effect group, and this needs to be improved)'. Among these, SH3 and SH10 should be taken into deeper consideration. This MCDM method is comprehensive and applicable to all companies facing problems that require group decision-making in a fuzzy environment. However, this study contains

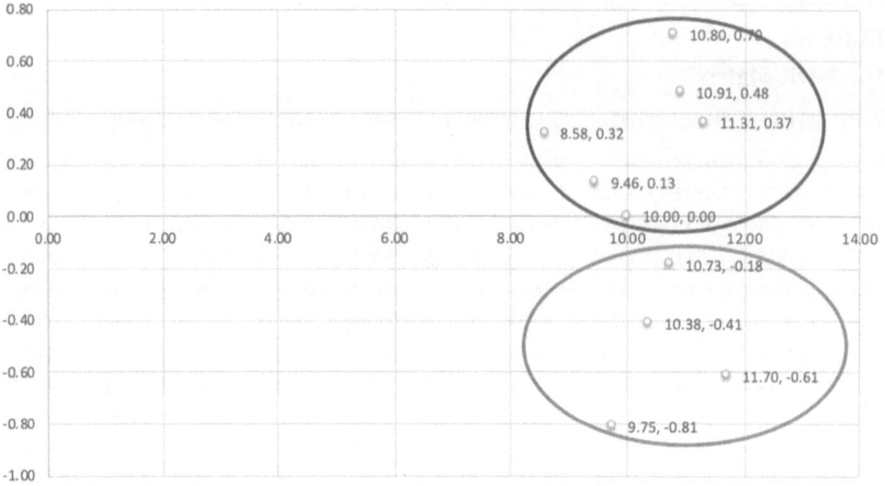

FIGURE 6.1 IRD Diagram.

TABLE 6.6
Cause/Effect Factors for Sustainable HRM Factors

	Ri	Ci	Ri + Ci	Ri − Ci	Cause/Effect
SH1	5.7507	5.0461	10.80	0.70	Cause
SH2	5.8385	5.4713	11.31	0.37	Cause
SH3	4.4728	5.2806	9.75	−0.81	Effect
SH4	5.0024	4.9987	10.00	0.00	Cause
SH5	4.9826	5.3928	10.38	−0.41	Effect
SH6	5.6960	5.2138	10.91	0.48	Cause
SH7	4.7960	4.6631	9.46	0.13	Cause
SH8	5.5448	6.1579	11.70	−0.61	Effect
SH9	4.4531	4.1289	8.58	0.32	Cause
SH10	5.2728	5.4565	10.73	−0.18	Effect

certain limitations. Future studies should try to involve empirical analysis to increase the implementation of sustainable supply chains through sustainable human resource practices. This would pave way for new trends in multi-criteria decision-making in the future.

REFERENCES

Ab Talib, M.S., & Hamid, A.B.A. (2014). Application of critical success factors in supply chain management. *International Journal of Supply Chain Management*, *3*(1), 21–33.

Al-Odeh, M., & Smallwood, J. 2012. Sustainable supply chain management: Literature review, trends, and framework. *IJCEM International Journal of Computational Engineering & Management ISSN*, *15*(1), 2230–7893.

Agarwal, V., Govindan, K., Darbari, J. D., & Jha, P. C. (2016). An optimization model for sustainable solutions towards implementation of reverse logistics under collaborative framework. *International Journal of System Assurance Engineering and Management*, *7*(4), 480–487.

Anastasiou, S. (2012). "Critical Human Resources Management Functions for Efficient Logistics and Supply Chain Management Education-Human Resources Management View Project Critical Human Resources Management Functions for Efficient Logistics and Supply Chain Management." *International Conference on Supply Chains*, no. July 2012.

Das, D. (2018). Sustainable supply chain management in Indian organisations: An empirical investigation. *International Journal of Production Research*, *56*(17), 5776–5794.

De Prins, P., Van Beirendonck, L., De Vos, A., & Segers, J. (2014). Sustainable HRM: Bridging theory and practice through the 'respect openness continuity (ROC)'-model. *Management Revue*, *25*(4), 263–284.

Docherty, P., Forslin, J., & Shani, A. B. (Eds). (2002). *Creating Sustainable Work Systems: Emerging Perspectives and Practice.* Psychology Press, London.

Ehnert, I., Harry, W., & Zink, K.J. (Eds). (2014). Sustainability and human resource management - Developing Sustainable Business Organizations. In *CSR, Sustainability, Ethics & Governance*. Vol. 6 (pp. 1–423). Springer.

Florescu, M.S., Ceptureanu, E.G., Cruceru, A.F., & Ceptureanu, S.I. (2019). Sustainable supply chain management strategy influence on supply chain management functions in the oil and gas distribution industry. *Energies*, *12*(9), 1632.

Gaur, L., Agarwal, V., & Anshu, K. (2020). Fuzzy DEMATEL approach to identify the factors influencing efficiency of indian retail websites. In *Strategic System Assurance and Business Analytics* (pp.?69–84). Springer, Singapore.

Govindan, K., Kaliyan, M., Kannan, D., & Haq, A. N. (2014). Barriers analysis for green supply chain management implementation in Indian industries using analytic hierarchy process. *International Journal of Production Economics*, *147*(PART B), 555–568.

Hohenstein, N.O., Feisel E., & Hartmann E. (2014). Human resource management issues in supply chain management research: A systematic literature review from 1998 to 2014. *International Journal of Physical Distribution and Logistics Management*, *44*(6), 434–463.

Khodakarami, M., Shabani, A., Saen, R. F., & Azadi, M. (2015). Developing distinctive two-stage data envelopment analysis models: An application in evaluating the sustainability of supply chain management. *Measurement*, 70, 62–74.

Kumar, A., Mangla, S.K., Luthra, S., & Ishizaka, A. (2019). "Evaluating the Human Resource Related Soft Dimensions in Green Supply Chain Management Implementation." *Production Planning and Control*, 30(9), 699–715.

Kumar, R., Kumar, V., and Singh, S. (2017a). Modeling and analysis on supply chain characteristics using ISM technique. *Apeejay Journal of Management and Technology*, *12*(1 & 2), 21–30.

Kumar, R., Kumar, V., and Singh, S. (2017b). Work culture enablers: Hierarchical design for effectiveness & efficiency. *International Journal of Lean Enterprise Research (IJLER)*, *2*(3), 189–201.

Malhotra, S., & Singh, T. (2018). A study on impact of training programs in the IT industry. *Training & Development Journal*, *9*(1), 42–47.

Maley, J. & Kramar, R. (2015). Sustainable HRM in the context of global uncertainty: Its value for MNCs and impact on the global manager?. In EURAM 2015: 15th Annual Conference of the European Academy of Management(pp. 1–33) EURAM, Kozminski University.

Mittal, V.K., Sindhwani, R., Singh, P.L., Kalsariya, V., and Salroo, F. (2018). Evaluating significance of green manufacturing enablers using MOORA method for Indian manufacturing sector. In Proceedings of the *International Conference on Modern Research in Aerospace Engineering* (pp.?303–314). Springer, Singapore.

Mokhtar, M.F., Omar, B., Nor, N.H.M., Pauzi, N.F.M., Hassan, S., & Mohamed, W.A.W. (2016). Social and Economic Concern of Supply Chain Sustainability (SCS)., (). IOP Conference Series: Materials Science and Engineering. Vol. 160 (p.?012073). IOP Publishing.

Opricovic, S. & Tzeng, G. H. (2003). Fuzzy multicriteria model for postearthquake land-use planning. *Natural Hazards Review*, *4*(2), 59–64.

Sindhwani, R., Mittal, V.K., Singh, P.L., Kalsariya, V., & Salroo, F. (2018). Modelling and analysis of energy efficiency drivers by fuzzy ISM and fuzzy MICMAC approach. *International Journal of Productivity and Quality Management*, *25*(2), 225–244.

Sindhwani, R., Singh, P.L., Iqbal, A., Prajapati, D.K., & Mittal, V.K. (2019). Modeling and analysis of factors influencing agility in healthcare organizations: An ISM approach. In K. Shanker, R. Shankar, & R. Sindhwani (Eds.) *Advances in Industrial and Production Engineering* (pp.?683–696). Springer, Singapore.

Slaper, T. F., & Hall T. J. (2011). The triple bottom line: What is it and how does it work. *Indiana Business Review,* 86(1), 4–8.

Sriyogi, K., Agrawal, R., & Sharma, V. (2017). Sustainable Supply Chain Management Practices in Indian Manufacturing Firms a Case-Based Research Proceedings of 3rd International Conference on Business analytics at Great Lakes Institute of Management, Chennai On December 24, 2013, Available at SSRN: https://ssrn.com/abstract=3034

Teixeira, A.A., Jabbour, C.J.C., De Sousa Jabbour, A.B.L., Latan, H., & De Oliveira, J.H.C. (2016). Green training and green supply chain management: Evidence from Brazilian firms. *Journal of Cleaner Production*, *116*(2016), 170–176.

Vimal, K.E.K., Mathiyazhagan, K., Agarwal, V., Luthra, S., & Sivakumar, K. (2020). Analysis of barriers that impede the elimination of single-use plastic in developing economy context. *Journal of Cleaner Production*, *272*, 122629.

Whelan, T. & Fink, C. (2016). The Comprehensive Business Case for Sustainability, Harward Business Review. Available on: https://hbr.org/2016/10/the-comprehensive-business-case-for-sustainability

Wirtenberg, J., Harmon, J., Russell, W.G., & Fairfield, K.D. (2007). HR's role in building a sustainable enterprise: Insights from some of the world's best companies. *Human Resource Planning*, *30*(1), 10–20.

Zaid, A.A., Jaaron, A.A.M., & Bon, A.T. (2018). The impact of green human resource management and green supply chain management practices on sustainable performance: An empirical study. *Journal of Cleaner Production*, *204*, 965–979.

7 A Novel Entropy Measure for Linguistic Intuitionistic Fuzzy Sets and Their Application in Decision-Making

Kamal Kumar, Naveen Mani, Amit Sharma, and Reeta Bhardwaj

CONTENTS

7.1 Introduction 121
7.2 Literature Review 122
    7.2.1 Preliminaries 123
    7.2.2 A Novel Entropy Measure for LIFS 125
7.3 MCDM Approaches Based on the Proposed EM 129
7.4 Comparative Study 133
7.5 Conclusion 134
References 134

7.1 INTRODUCTION

In reality, the first big challenge for decision-maker(s) (DMks) in solving MCDM issues is to determine the environment to the give the assessments for performance towards the criteria of the alternatives (Kumar et al., 2019). This aspect makes the problem more complex and uncertain for decision-makers (DMks) and cannot provide their assessments in the form of crisp numbers. To remove such types of difficulties for the DMks, the 'fuzzy set' (FS) was introduced by Zadeh (1965), after which, extensions such as 'intuitionistic FS' (IFS) (Atanassov, 1986) and 'interval-valued IFS' (IVIFS) (Atanassov & Gargov, 1989) have been proposed as powerful tools in handling levels of uncertainty. These theories have been applied widely in the field of MCDM. This is similar to how Garg and Kumar (2017, 2020) defined the correlation coefficient and novel exponential distance measure based on the SPA theory. Gupta et al. (2018) determined the TOPSIS method for IVIFSs. The TOPSIS system, based on the SPA principle, was introduced by Kumar and Garg (2018).

In MCDM issues, criteria weights play a crucial role during the aggregation process because fluctuation in the weights can change the RO of the alternatives (Mittal et al., 2019). In certain MCDM problems, criteria weights are provided by the DMks, but in some situations, DMks cannot set the criteria weights due to the fuzziness of the data. To handle this, entropy measure (EM) is an effective tool, which is generally known as the knowledge measure derived from 'The Mathematical Theory of Communication' (Shannon, 1948). Therefore, the concept of fuzzy EM was introduced by Deluca, (1972). Later on, Szmidt and Kacprzyk (2001) extended this concept for IFSs. Zhang and Jiang (2008) generalized the logarithmic EM of DeLuca & Termini (1993) for IFSs.

All the above studies of MCDM are based on quantitative aspects (Sindhwani et al., 2018). In some real-life MCDM issues, a decision must be made, but the DMk assessments may not be in the form of numerical values (NVs). However, DMks can describe their evaluations in the form of qualitative aspects. For instance, to examine the home's location, DMks may use the linguistic variables (LVs) 'good,' 'average,' and 'bad' in place of NVs. In handling such types of cases, firstly, the concept of LVs was introduced by Zadeh (1975). Afterwards, various researchers (Dong et al., 2010; Herrera & Martinez, 2001) applied LVs into decision theories. Xu (2004) introduced the AOs for LVs. Xu et al. (2012) proposed the LVs power aggregation operators (AOs) with applications in MCDM issues. After that, Zhang (2014) defined linguistic IFS (LIFS) for simpler expressions of qualitative assessment where 'membership' and 'non-membership' grades are in LVs. AOs for the aggregation of linguistic intuitionistic fuzzy values (LIFVs) was suggested by Chen et al. (2015).

However, in the literature survey and under the LIFS context, it was discovered that the EM for the LIFS had not been investigated. Keeping under consideration that, in this chapter, the authors construct an EM to depict the fuzziness of LIFS. The legality and validity of certain features of the proposed EM have been verified. The primary objective of defining the EM is to assess the weight of the parametres when these are either absolutely unknown or partially understood. To show the application of the proposed EM, the proponents introduce two MCDM approaches by considering attribute weight characteristics, which are either absolutely unknown or only partially known. Real-life illustrative examples have been studied to test the developed methods and are subsequently contrasted with the existing strategies to exhibit the interests of the proposed strategy.

This chapter is organized as follows: A literature review that is related to MCDM is provided in Section 2. In Section 3, some basic concepts of LIFSs are briefly introduced. In Section 4, an EM has been constructed for LIFS, and certain properties and axiom terminology are also defined. Section 5 describes two MCDM approaches along with their applications. Ultimately, the chapter is concluded in Section 6.

7.2 LITERATURE REVIEW

Recently, fuzzy theory and its extensions are widely used in different disciplines, such as computation intelligence, decision-making (DM), reliability analysis *etc.*

Recently in the DM field, these theories have been applied more by researchers (Garg & Kumar, 2017, 2019, 2020; Gupta et al., 2018; Jiang & Wei, 2018; Liu & Li, 2018) to solve real-world MCDM issues. An MCDM process consists mainly of two phases. The first is designing a suitable mechanism that aggregates the different DMk's preferences into a collective one. Afterward, the second is creating an effective measure for the ROs of alternatives. AO is an important tool for the former portion. Under the IFS and IVIFS context, several types of AOs have been developed by researchers in order to fuse information. Similarly, Xu & Yager (2006) and Xu (2007a) introduced the geometric and averaging AOs respectively for aggregating the IFS environment. Thereupon, Xu and Chen (2007) and Xu (2007b) extended the AOs for IVIFS environment. Garg (2016a, 2016b) defined intuitionistic fuzzy interactive averaging (IFIA) AOs by adding the hesitance degree into these operators. A far-reaching review of the diverse methodologies under the IFSs and additionally IVIFSs are summarized by Xu and Zhao (2016). Recently, Deschrijver and Kerre (2002) introduced the generalized union and intersection for IFSs. Afterwards, authors Wang and Liu (2011, 2012) presented various Einstein operators for aggregating the IFSs.

Wei et al. (2012) presented a new EM by using the cosine function for IFS. Wang and Wang (2012) defined the cotangent EM for IFS. Verma and Sharma (2013) presented the exponential EM of IFSs. Furthermore, all of the previously existing EM of IFSs do not contain the degree of hesitancy of IFS. Liu and Ren (2014) realized some drawbacks of the existing EMs of IFS, as explained above, and presented an EM by containing the degree of hesitance of IFS. Garg and Kaur (2018) proposed a novel (R, S) Norm EM for IFS. Another main important part of solving MCDM issues is aggregating the provided DMks data.

For the qualitative aspect, Peng et al. (2018) developed the AOs based on Frank operations for LIFVs. Various improved operative legislation for LIFVs and AOs based on this was given by Liu and Wang (2017). Garg and Kumar (2018b) proposed AOs for aggregating linguistic connection numbers and the DM method under the LIFVs environment based on the set pair analysis. Xian et al. (2017) recommended new hybrid AOs for LIFVs as well as a DM approach based on this. Liu and Qin (2017) put PA AOs for solving DM issues under the LIFVNs environment. Liu and Liu (2017) proposed power BM AOs to fuse LIFVs. Garg and Kumar (2018a) defined possibility degree and Einstein AOs for LIFVs environment.

7.2.1 Preliminaries

A few basic principles related to the theory of LIFS are characterized in this section.

Definition 7.2.1: (Herrera & Martinez, 2001) 'Let $S = \{s_t \mid t = 0, 1, 2, \ldots, h\}$ be an LV set of finite odd cardinality, where s_t represents a possible value for a LV, and h is the positive integer'.

In such cases, s_t must be satisfy the following properties (Herrera & Martinez, 2001):

$$(i)\quad s_k \leq s_t \Leftrightarrow k \leq t$$
$$(ii)\quad \bar{s}_k = s_{h-k}$$
$$(iii)\quad \max(s_p, s_t) = s_t \Leftrightarrow s_t > s_p$$
$$(iv)\quad \min(s_p, s_t) = s_p \Leftrightarrow s_t > s_p$$

Afterward, Xu (2004) provided the concept of continuous LVs as:

$$S_{[0,h]} = \{s_z | s_0 \leq s_z \leq s_h\}.$$

Definition 7.2.2: (Zhang, 2014) A LIFS 'H' in the finite universe set X is defined as:

$$H = \{\langle x, s_{\tau_H(x)}, s_{\theta_H(x)}\rangle | x \in X\}, \quad (7.1)$$

Where $s_{\tau_H(x)}, s_{\theta_H(x)} \in S[0, h]$ characterizes the 'membership' and 'non-membership' grades of x to H correspondingly, such that $0 \leq \tau_H(x) + \theta_H(x) \leq h$ holds $\forall x$. The 'linguistic intuitionistic index' of x to H is defined as $s_{\pi_H(x)} = s_{h-s_{\tau_H(x)}-s_{\theta_H(x)}}$. Generally, the order couple $\langle s_\tau, s_\theta\rangle$ is called LIFVs.

Theorem 7.2.1: (Chen et al., 2015) For two LIFSs $H_1 = \{(x, s_{\tau_1(x)}, s_{\theta_1(x)}) | x \in X\}$ and $H_2 = \{(x, s_{\tau_2(x)}, s_{\theta_2(x)}) | x \in X\}$, we have the following:

a. $H1 = H2 \Leftrightarrow \tau_1(x) = \tau_2(x)$ and $\theta_1(x) = \theta_2(x)$;
b. $H1 \subseteq H2$ if $\tau_1(x) \leq \tau_2(x)$ and $\theta_1(x) \geq \theta_2(x)$;
c. $H^c = \{\langle x, s_{\theta_H(x)}, s_{\tau_H(x)}\rangle | x \in X\}$.

Definition 7.2.3: (Chen et al., 2015) The score value $S(\gamma)$ for LIFV $\gamma = (s_\tau, s_\theta)$ is defined as:

$$S(\gamma) = \tau - \theta, \quad (7.2)$$

where $S(\gamma) \in [-h, h]$.

Definition 7.2.4: (Chen et al., 2015) The accuracy function H(γ) for LIFV $\gamma = (s_\tau, s_\theta)$ is defined as:

$$H(\gamma) = \tau + \theta \quad (7.3)$$

where $H(\gamma) \in [0, h]$.

Definition 7.2.5: (Chen et al., 2015) For an LIFS $H = \{\langle x, s_{\tau_H(x_t)}, s_{\theta_H(x_t)}\rangle | x_t \in X$ and any real $\lambda > 0$, defines the LIFS H^λ as follows:

$$H^{\lambda} = \left\{ \left\langle x,\ s_{h\left(\frac{\tau(x_t)}{h}\right)^{\gamma}}\ s_{h\left(1-\left(1-\frac{\theta(x_t)}{h}\right)^{\gamma}\right)} \right\rangle \mid x_t \in X \right\} \tag{7.4}$$

7.2.2 A Novel Entropy Measure for LIFS

We will present an entropy measure (EM) for LIFS in this segment. Let $\varphi(X)$ be the gathering of all of the LIFSs.

Definition 7.2.6: If $H \in \varphi(X)$, then EM $E : \varphi(X) \rightarrow [0,\ 1]$ fulfils the subsequent properties:

(P1) $E(H) = 0 \Leftrightarrow H$ is a linguistic set;

(P2) $E(H) = 1 \Leftrightarrow \tau_H(x) = \theta_H(x),\ \forall\, x \in X$;

(P3) $E(H) = E(H^c)$;

(P4) *If* $H_1 \leq H_2$ *then* $E(H_1) \leq E(H_2)$.

Definition 7.2.7: For an LIFS $H = \{\langle x, s_{\tau_H(x_t)}, s_{\theta_H(x_t)} \rangle | x_t \in X\}$, we define the following entropy measure:

$$E(H)\frac{1}{3nh}\sum_{t=1}^{n}\left(4\sqrt{\tau_H(x_t)\theta_H(x_t)} + \pi_H(x_t) + 2\sqrt{(h-\tau_H(x_t))(h-\theta_H(x_t))}\right) \tag{7.5}$$

Theorem 7.2.2: The EM $E(H)$ for LIFS H defined in Eq. (7.5) satisfies (P1)–(P4) properties as given in Definition 7.2.6.

Proof. Let $H = \{\langle x, s_{\tau_H(x_t)}, s_{\theta_H(x_t)} \rangle | x_t \in X\}$ be an IFS.

(P1) We have the following:

$$E(H) = 0$$

$$\Leftrightarrow \frac{1}{3nh}\sum_{t=1}^{n}\left(4\sqrt{\tau_H(x_t)\,\theta_H(x_t)} + \pi_H(x_t) + 2\sqrt{(h-\tau_H(x_t))(h-\theta_H(x_t))}\right) = 0$$

$$\Leftrightarrow \sqrt{\tau_H(x_t)\theta_H(x_t)} = 0,\ \pi_H(x_t) = 0 \quad \textit{and} \quad \sqrt{(h-\tau_H(x_t))(h-\theta_H(x_t))} = 0$$

$$\Leftrightarrow \pi_H(x_t) = 0,\ \tau_H(x_t)\ or\ \theta_H(x_t) = 0$$

$\Leftrightarrow$ H is a linguistic set.

(P2) Since $\sqrt{xy}$ achieves its maximum value $\frac{x+y}{2}$ when $x = y$; therefore, we have:

$$E(H) = 1$$

$$\Leftrightarrow \frac{1}{3nh}\sum_{t=1}^{n}\left(4\sqrt{\tau_H(x_t)\theta_H(x_t)} + \pi_H(x_t) + 2\sqrt{(h-\tau_H(x_t))(h-\theta_H(x_t))}\right) = 1$$

$$\Leftrightarrow 4\sqrt{\tau_H(x_t)\theta_H(x_t)} + \pi_H(x_t) + 2\sqrt{(h-\tau_H(x_t))(h-\theta_H(x_t))} = 3h$$

$$\Leftrightarrow \tau_H(x_t) = \theta_H(x_t)$$

(P3) Since $H^c = \left\{\left\langle x, s_{\theta_H(x_t)}, s_{\tau_H(x_t)}\right\rangle | x_t \in X\right\}$, we therefore have:

$$E(H) = \frac{1}{3nh}\sum_{t=1}^{n}\left(4\sqrt{\tau_H(x_t)\,\theta_H(x_t)} + \pi_H(x_t) + 2\sqrt{(h-\tau_H(x_t))(h-\theta_H(x_t))}\right)$$
$$= \frac{1}{3nh}\sum_{t=1}^{n}\left(4\sqrt{\theta_H(x_t)\tau_H(x_t)} + \pi_H(x_t) + 2\sqrt{(h-\theta_H(x_t))(h-\tau_H(x_t))}\right)$$
$$= E(H^c)$$

(P4) Construct the function $f(x, y) = 4\sqrt{xy} + h - x - y + 2\sqrt{(h-x)(h-y)}$, where, $x, y \in [0, h]$ and $x + y \leq h$. At this point, we will prove that the function $f(x, y)$ is increasing with x and decreasing with y when $x \leq y$.

The partial derivatives of $f(x, y)$ with regard to x and y can be derived as:

$$\frac{\partial f(x, y)}{\partial x} = 2\sqrt{\frac{y}{x} - \sqrt{\frac{h-y}{h-x}} - 1}$$

$$\frac{\partial f(x, y)}{\partial y} = 2\sqrt{\frac{x}{y} - \sqrt{\frac{h-x}{h-y}} - 1}$$

Since $\frac{\partial f(x,y)}{\partial x} \leq 0$, $\frac{\partial f(x,y)}{\partial x} \leq 0$, when $x \leq y$; therefore, $f(x, y)$ is increasing with x and decreasing with y. Thus, $f(\tau_1(x_t), \theta_1(x_t)) \leq f(\tau_2(x_t), \theta_2(x_t))$, when $\tau_2(x_t) \leq \theta_2(x_t)$ and $\tau_1(x_t) \leq \tau_2(x_t)$, $\theta_1(x_t) \geq \theta_2(x_t)$.

Similarly, $\frac{\partial f(x,y)}{\partial x} \le 0$, $\frac{\partial f(x,y)}{\partial x} \ge 0$, when $x \le y$; therefore, $f(x, y)$ is decreasing with x and increasing with y. Thus, $f(\tau_1(x_t), \theta_1(x_t)) \le f(\tau 2(x_t), \theta_2(x_t))$ when $\tau 2(x_t) \ge \theta_2(x_t)$ and $\tau_1(x_t) \ge \tau_2(x_t)$, $\theta_1(x_t) \le \theta_2(x_t)$.

Therefore, If $H_1 \preccurlyeq H_2$, then $\frac{1}{3n}\sum_{t=1}^{n} f(\tau_1(x_t), \theta_1(x_t)) \le \frac{1}{3n}\sum_{t=1}^{n} f(\tau_2(x_t), \theta_2(x_t))$. Hence $E(H_1) \le E(H_2)$.

Example 7.2.1: Consider an LIFS H 'LARGE' over X as follows:

$$H = \{(x_1, s_1, s_7), (x_2, s_2, s_4), (x_3, s_5, s_1), (x_4, s_7, s_0), (x_5, s_8, s_0)\} \quad (7.6)$$

By utilizing Eq. (7.4), we generate the LIFSs $H^{1/2}$('Less LARGE'), H^2 ('Very LARGE'), H^3 ('Quit very LARGE') and H^4 ('Very, very LARGE') as follows:

$$H^{1/2} = \left\{ \begin{array}{c} \langle x_1, s_{2.8284}, s_{5.1716}\rangle, \langle x_2, s_{4.0000}, s_{2.3431}\rangle, \langle x_3, s_{6.3246}, s_{0.5167}\rangle, \\ \langle x_4, s_{7.4833}, s_{s0}\rangle, \langle x_5, s_{8.0000}, s_0\rangle \end{array} \right\}$$

$$H^2 = \left\{ \begin{array}{c} \langle x_1, s_{2.8284}, s_{5.1716}\rangle, \langle x_2, s_{4.0000}, s_{2.3431}\rangle, \langle x_3, s_{6.3246}, s_{0.5167}\rangle, \\ \langle x_4, s_{7.4833}, s_0\rangle, \langle x_5, s_{8.0000}, s_0\rangle \end{array} \right\}$$

$$H^3 = \left\{ \begin{array}{c} \langle x_1, s_{0.0156}, s_{7.9844}\rangle, \langle x_2, s_{0.1250}, s_{7.0000}\rangle, \langle x_3, s_{1.9531}, s_{2.6406}\rangle, \\ \langle x_4, s_{5.3594}, s_0\rangle, \langle x_5, s_{8.0000}, s_0\rangle \end{array} \right\}$$

$$H^4 = \left\{ \begin{array}{c} \langle x_1, s_{0.0020}, s_{7.9980}\rangle, \langle x_2, s_{0.0313}, s_{7.5000}\rangle, \langle x_3, s_{1.2207}, s_{3.3105}\rangle, \\ \langle x_4, s_{4.6895}, s_0\rangle, \langle x_5, s_{8.0000}, s_0\rangle \end{array} \right\}$$

At this point, we will calculate the EM by utilizing Eq. (7.5) for the above-defined LIFSs as follows:

$$E(H^{\frac{1}{2}}) = 0.5535,\ E(H) = 0.5479,\ E(H^2) = 0.4613,\ E(H^3) = 0.4005,\ E(H^4) = 0.3628$$

Thus, the proposed EM satisfies an effective pattern $E(H^{\frac{1}{2}}) > E(H) > E(H^2) > E(H^3) > E(H^4)$. Hence, this is a valid EM.

Theorem 7.2.3: Let $H_1 = \{\langle x, s_{\tau_1(x_t)}, s_{\theta_1(x_t)}\rangle \mid x_t \in X\}$ and $H_2 = \{\langle x, s_{\tau_2(x_t)}, s_{\theta_2(x_t)}\rangle \mid x_t \in X\}$ be two LIFSs, such that either $H_1 \subseteq H_2$ or $H_1 \supseteq H_2 \forall x_t \in X$, then:

$$E(H_1 \cup H_2) + E(H_1 \cap H_2) = E(H_1) + E(H_2)$$

Proof. Consider two set X_1 and X_2 of X, such that:

$$X_1 = \{x_t \in X \mid H_1 \subseteq H_2\}, \quad X_2 = \{x_t \in X \mid H_1 \supseteq H_2\}$$

i.e. $\forall x_t \in X_1$, we have $\tau_1(x_t) \leq \tau_2(x_t)$, $\theta_1(x_t) \geq \theta_2(x_t)$ *and* $\forall x_t \in X_2$, $\tau_1(x_t) \geq \tau_2(x_t)$, $\theta_1(x_t) \leq \theta_2(x)$.

Therefore, according to the proposed EM, we have:

$$E(H_1 \cup H_2) = \frac{1}{3nh} \sum_{x_t \in X} \begin{pmatrix} 4\sqrt{\tau_{H_1 \cup H_2}(x_t)\theta_{H_1 \cup H_2}(x_t)} + \pi_{H_1 \cup H_2}(x_t) \\ +2\sqrt{(h - \tau_{H_1 \cup H_2}(x_t))(h - \theta_{H_1 \cup H_2}(x_t))} \end{pmatrix}$$

$$= \frac{1}{3nh} \begin{bmatrix} \sum_{x_t \in X_1} (4\sqrt{\tau_{H_2}(x_t)\theta_{H_2}(x_t)} + \pi_{H_2}(x_t) \\ + 2\sqrt{(h - \tau_{H_2}(x_t))(h - \theta_{H_2}(x_t))}) \\ + \sum_{x_t \in X_2} (4\sqrt{\tau_{H_1}(x_t)\theta_{H_1}(x_t)} + \pi_{H_1}(x_t) \\ + 2\sqrt{(h - \tau_{H_1}(x_t))(h - \theta_{H_1}(x_t))}) \end{bmatrix}$$

Similarly,

$$E(H_1 \cap H_2) = \frac{1}{3nh} \begin{bmatrix} \sum_{x_t \in X_1} (4\sqrt{\tau_{H_1}(x_t)\theta_{H_1}(x_t)} + \pi_{H_1}(x_t) \\ + 2\sqrt{(h - \tau_{H_1}(x_t))(h - \theta_{H_1}(x_t))}) \\ + \sum_{x_t \in X_2} (4\sqrt{\tau_{H_2}(x_t)\theta_{H_2}(x_t)} + \pi_{H_2}(x_t) + \pi_{H_2}(x_t) \\ + 2\sqrt{(h - \tau_{H_2}(x_t))(h - \theta_{H_2}(x_t))}) \end{bmatrix}$$

$$E(H_1 \cup H_2) + E(H_1 \cap H_2) = \frac{1}{3nh} \begin{bmatrix} \sum_{x_t \in X_1} (4\sqrt{\tau_{H_2}(x_t)\theta_{H_2}(x_t)} + \pi_{H_2}(x_t) \\ + 2\sqrt{(h - \tau_{H_2}(x_t))(h - \theta_{H_2}(x_t))}) \\ + \sum_{x_t \in X_2} (4\sqrt{\tau_{H_1}(x_t)\theta_{H_1}(x_t)} + \pi_{H_1}(x_t) \\ + 2\sqrt{(h - \tau_{H_1}(x_t))(h - \theta_{H_1}(x_t))}) \end{bmatrix}$$

$$+ \frac{1}{3nh} \begin{bmatrix} \sum_{x_t \in X_1} (4\sqrt{\tau_{H_1}(x_t)\theta_{H_1}(x_t)} + \pi_{H_1}(x_t) \\ + 2\sqrt{(h - \tau_{H_1}(x_t))(h - \theta_{H_1}(x_t))}) \\ + \sum_{x_t \in X_2} (4\sqrt{\tau_{H_2}(x_t)\theta_{H_2}(x_t)} + \pi_{H_2}(x_t) + \pi_{H_2}(x_t) \\ + 2\sqrt{(h - \tau_{H_2}(x_t))(h - \theta_{H_2}(x_t))}) \end{bmatrix}$$

$$= E(H_1) + E(H_2)$$

7.3 MCDM APPROACHES BASED ON THE PROPOSED EM

In this segment, the two MCDM approaches are constructed in the sense of LIFVs on the basis of the proposed EM.

Approach I: *'When weights of criteria are partially known'*

Adopt *'m'* alternatives $H = \{H_1, H_2, \ldots, H_m\}$ and *'n'* different criterion $G = \{G_1, G_2, \ldots, G_n\}$ with partial known weight $\omega_t \in [\omega_{t,}^l\ \omega_t^u]$ of the criterion G_t. Assume that the DMk assesses attribute G_t of alternative A_k using an LIFV $\tilde{\gamma}_{kt} = \langle s_{\tilde{\tau}_{kt}}, s_{\tilde{\theta}_{kt}} \rangle$, $k = 1, 2, \ldots, m$, and $t = 1, 2, \ldots, n$, to construct the decision matrix (DMx) $\tilde{R} = (\tilde{\gamma}_{kt})_{m \times n}$, shown as follows:

$$\tilde{R} = \begin{array}{c} \\ A_1 \\ A_2 \\ \vdots \\ A_m \end{array} \begin{array}{c} \begin{array}{cccc} G_1 & G_2 & \ldots & G_n \end{array} \\ \begin{pmatrix} \tilde{\gamma}_{11} & \tilde{\gamma}_{12} & \cdots & \tilde{\gamma}_{1n} \\ \tilde{\gamma}_{21} & \tilde{\gamma}_{22} & \cdots & \tilde{\gamma}_{2n} \\ \vdots & \vdots & \ddots & \vdots \\ \tilde{\gamma}_{m1} & \tilde{\gamma}_{m2} & \cdots & \tilde{\gamma}_{mn} \end{pmatrix} \end{array}$$

The procedure of the MCDM proposed method includes the following steps:

Step 1: By using Eq. (7.7), convert the DMx $\tilde{R} = (\tilde{\gamma}_{kt})_{m \times n}$ into the normalized DMx $R = (\gamma_{kt})_{m \times n} = (\langle s_{\tau_{kt}}, s_{\theta_{kt}} \rangle)_{m \times n}$ as follows:

$$\gamma_{kt} = \begin{cases} \langle s_{\tilde{\tau}_{kt}}, s_{\tilde{\theta}_{kt}} \rangle, & \text{if } G_t \text{ is a benifit type attribute} \\ \langle s_{\tilde{\theta}_{kt}}, s_{\tilde{\tau}_{kt}} \rangle, & \text{if } G_t \text{ is a cost type attribute} \end{cases} \tag{7.7}$$

Step 2: By using the Eq. (7.5), calculate the EM $E(\gamma_{kt})$ for each LIFV γ_{kt} given in the DMx $R = (\gamma_{kt})_{m \times n}$ to obtain the EM matrix $D_E = (E(\gamma_{kt}))_{m \times n}$, where $k = 1, 2, m$ and $t = 1, 2,, n$.

Step 3: Based on the EM matrix $D_E = (E(\gamma_{kt}))_{m \times n}$, we construct an LPP to obtain the weights of criteria $\omega_1, \omega_2, \ldots, \omega_n$ as follows:

$$\sum_{t=1}^{n} \sum_{k=1}^{m} E(\gamma_{kt})\omega_{kt}$$

s.t.

$$\omega_t^l \le \omega_t \le \omega_t^u,$$

$$\sum_{t=1}^{n} \omega_t = 1.$$

Determine the optimal criterion's weight by solving the above formulated LPP.

Step 4: Compute the overall performance $Q(H_k) = s_{z_k}$ of alternative $H_k (k = 1, 2, ..., m)$, where z_k is defined as

$$z_k = \begin{cases} \sum_{t=1}^{n} w_t \frac{(t_{kt} - \theta_{kt})(t_{kt} + \theta_{kt})}{h} & \text{if } t_{kt} \neq \theta_{kt} \\ \sum_{t=1}^{n} w_t \frac{t_{kt}(2h - t_{kt} - \theta_{kt})}{h} & \text{if } t_{kt} = \theta_{kt} \end{cases} \tag{7.8}$$

Step 5: Calculate the RO of the alternatives $H_1, H_2, ..., H_m$ by the descending sequence of overall performance s_{z_k}, $k = 1, 2, ..., m$.

In the following, we adopt a real-life MCDM example from (Zhang et al., 2017) to evaluate the proposed MCDM Approach I.

Example 7.2.2: A mining office investigates the safety conditions of four coal mines, denoted by H_1, H_2, H_3, and H_4 in a given region. Based on the analysis, the protection climates affected by several variables are taken into account, and the following four requirements are listed with weights ω_1, ω_2, ω_3, and ω_4 respectively.

i. G_1 ('technological equipment')
ii. G_2 ('geological condition')
iii. G_3 ('human diathesis')
iv. G_4 ('management level')

After a heated conversation, incomplete information about the criterion weight was provided as $0.16 \le \omega_1 \le 0.30$, $0.20 \le \omega_2 \le 0.30$, $0.15 \le \omega_3 \le 0.35$, $0.18 \le \omega_4 \le 0.28$. Experts must use their technical expertise and work experience according to LVs to correctly analyze the four coal mines as s_0 = 'extremely poor,' s_1 = 'very poor,' s_2 = 'poor,' s_3 = 'slightly poor,' s_4 = 'fair,' s_5 = 'slightly good,' s_6 = 'good,' s_7 = 'very good,' and s_8 = 'extremely good' under the given criteria. The assessment of the decision-maker by using LIFVs and are given in DMx as follows:

$$\tilde{R} = \begin{array}{c} \\ H_1 \\ H_2 \\ H_3 \\ H_4 \end{array} \begin{array}{c} \begin{array}{cccc} G_1 & G_2 & G_3 & G_4 \end{array} \\ \begin{pmatrix} \langle s_5, s_1 \rangle & \langle s_4, s_3 \rangle & \langle s_6, s_1 \rangle & \langle s_4, s_2 \rangle \\ \langle s_4, s_2 \rangle & \langle s_6, s_1 \rangle & \langle s_4, s_2 \rangle & \langle s_5, s_3 \rangle \\ \langle s_6, s_1 \rangle & \langle s_4, s_2 \rangle & \langle s_5, s_3 \rangle & \langle s_5, s_1 \rangle \\ \langle s_5, s_2 \rangle & \langle s_3, s_3 \rangle & \langle s_7, s_1 \rangle & \langle s_6, s_1 \rangle \end{pmatrix} \end{array}$$

Step 1: There is no need for the normalizing process due to all benefit type of criteria. Hence, we have = $(\gamma_{kt})_{4\times4} = (\tilde{\gamma}_{kt})_{4\times4}$.

Step 2: By using Eq. (7.5), calculate EM $E(\gamma_{kt})$ for each LIFV given in the DMx $R = (\gamma_{kt})_{4\times4}$ to obtain the EM matrix $D_E = (E(\gamma_{kt}))_{4\times4}$, where $k = 1, 2, 3, 4$, and $t = 1, 2, 3, 4$.

$$D_E = \begin{pmatrix} 0.8379 & 0.9917 & 0.7617 & 0.9630 \\ 0.9630 & 0.7617 & 0.9630 & 0.9682 \\ 0.7617 & 0.9630 & 0.9682 & 0.8379 \\ 0.9223 & 1.0000 & 0.6614 & 0.7617 \end{pmatrix}$$

Step 3: Based on the EM matrix $D_E = (E(\gamma_{kt}))_{4\times4}$, we construct the following LPP:

$$max\ 3.4849\omega_1 + 3.7164\omega_2 + 3.3544\omega_3 + 3.5308\omega_4$$

s.t.

$$0.16 \le \omega_1 \le 0.30,$$

$$0.20 \le \omega_2 \le 0.30,$$

$$0.15 \le \omega_3 \le 0.35,$$

$$0.18 \le \omega_4 \le 0.28,$$

$$\omega_1 + \omega_2 + \omega_3 + \omega_4 = 1.$$

Upon solving the LPP above, we obtain $\omega_1 = 0.27$, $\omega_2 = 0.30$, $\omega_3 = 0.15$, *and* $\omega_4 = 0.28$.

Step 4: For the alternatives H_k, $(k = 1, 2, 3, 4)$, the measure the overall performance $Q(H_k) = s_{z_k}$ and obtained values are given as:

$$Q(H_1) = s_{2.1488},\ Q(H_2) = s_{2.5025},\ Q(H_3) = s_{2.7713},\ Q(H_4) = s_{3.9587}$$

Step 5: Since $Q(H_4) > Q(H_3) > Q(H_2) > Q(H_1)$, therefore $H_4 \succ H_3 \succ H_2 \succ H_1$. Thus, the best option is H_4 for this assignment.

Approach II: *'When weights of criteria are absolutely unknown'*

In this case, an MCDM approach, where the weights of parametres are absolutely unknown, is developed.

Adopt 'm' alternatives $H = \{H_1, H_2, \ldots, H_m\}$ and n different criteria $G = \{G_1, G_2, \ldots, G_n\}$. DMks evaluate the alternatives H_k towards the criterion G_t by using the LIFV $\tilde{\gamma}_{kt} = \langle s_{\tilde{\tau}_{kt}}, s_{\tilde{\theta}_{kt}} \rangle$, $k = 1, 2, \ldots, m$, and $t = 1, 2, \ldots, n$, to construct the DMx $\tilde{R} = (\tilde{\gamma}_{kt})_{m\times n}$ shown as follows:

$$\tilde{R} = \begin{array}{c} \\ A_1 \\ A_2 \\ \vdots \\ A_m \end{array} \begin{array}{c} \begin{array}{cccc} G_1 & G_2 & \ldots & G_n \end{array} \\ \begin{pmatrix} \tilde{\gamma}_{11} & \tilde{\gamma}_{12} & \cdots & \tilde{\gamma}_{1n} \\ \tilde{\gamma}_{21} & \tilde{\gamma}_{22} & \cdots & \tilde{\gamma}_{2n} \\ \vdots & \vdots & \ddots & \vdots \\ \tilde{\gamma}_{m1} & \tilde{\gamma}_{m2} & \cdots & \tilde{\gamma}_{mn} \end{pmatrix} \end{array}$$

This MCDM proposed method includes the following steps:

Step 1: Follow Approach I Step 1.

Step 2: By using Eq. (7.5), calculate the EM $E(\gamma_{kt})$ for each LIFV given in the DMx $R = (\gamma_{kt})_{m\times n}$ to obtain the EM matrix $D_E = (E(\gamma_{kt}))_{m\times n}$, *where* $k = 1, 2, .., m$ *and* $t = 1, 2, .., n$.

Step 3: Compute the weight of the criterion G_t, $t = 1, 2, \ldots, n$, as:

$$\omega_t = \frac{1 - e_t}{n - \sum_{t=n}^{n} e_t}$$

Where $e_t = \frac{1}{m}\sum_{k=1}^{m} E(\gamma_{kt})$ and $E(\gamma_{kt}) = \frac{1}{3}(4\sqrt{\tau_{kt}\theta_{kt}} + \pi_{kt} + 2\sqrt{(1 - \theta_{kt})(1 - \theta_{kt})})$ is the entropy measure for $\gamma_{kt} = \langle \tau_{kt}, \theta_{kt} \rangle$.

Step 4: Compute the overall performance $Q(H_k) = s_{z_k}$ of each alternative $H_k (k = 1, 2, \ldots, m)$, where z_k is defined as follows:

$$z_k = \begin{cases} \sum_{t=1}^{n} w_t \frac{(t_{kt} - \theta_{kt})(t_{kt} + \theta_{kt})}{h} & \text{if } t_{kt} \neq \theta_{kt} \\ \sum_{t=1}^{n} w_t \frac{t_{kt}(2h - t_{kt} - \theta_{kt})}{h} & \text{if } t_{kt} = \theta_{kt} \end{cases}$$

Step 5: Calculate the RO of the alternatives $H_1, H_2, \ldots, H_m$ according to the descending sequence of overall performance s_{z_k}, where $k = 1, 2, \ldots, m$.

Example 7.2.3: We consider an illustration of MCDM in this case, which was mentioned and defined in Example 7.2.1, where the four alternatives H_1, H_2, H_3

and H_4 are evaluated by using the LIFVs towards the criteria G_1, G_2, G_3 and G_4 with absolutely unknown weights. To handle the previously mentioned MCDM issue, we utilize the proposed MCDM Approach II as follows:

Step 1: There is no need for normalizing due to all benefit-type of criteria.
Step 2: By using Eq. (7.5), calculate the EM $E(\gamma_{kt})$ for each LIFN given in the DMx $R = (\gamma_{kt})_{4\times4}$ to obtain the EM matrix $D_E = (E(\gamma_{kt}))_{4\times4}$, where $k = 1, 2, 4$ and $t = 1, 2, 4$.

$$D_E = \begin{pmatrix} 0.8379 & 0.9917 & 0.7617 & 0.9630 \\ 0.9630 & 0.7617 & 0.9630 & 0.9682 \\ 0.7617 & 0.9630 & 0.9682 & 0.8379 \\ 0.9223 & 1.0000 & 0.6614 & 0.7617 \end{pmatrix}$$

Step 3: Based on the EM matrix $D_E = (E(\gamma_{kt}))_{4\times4}$, we obtain $\omega_1 = 0.2692$, $\omega_2 = 0.1482$, $\omega_3 = 0.3374$, and $\omega_4 = 0.2452$.
Step 4: Calculate the overall performance $Q(H_k) = s_{z_k}$, $k = 1, 2, 3, 4$, for the alternatives H_1, H_2, H_3, and H_4, and obtained values are given as:

$$Q(H1) = s2.7812, \;\; Q(H2) = s2.0487, \;\; Q(H3) = s2.8105, \;\; Q(H4) = s4.3595$$

Step 5: Since $Q(H_4) > Q(H_3) > Q(H_1) > Q(H_2)$; therefore $H_4 > H_3 > H_1 > H_2$, and hence, H_4 is the best alternative.

7.4 COMPARATIVE STUDY

To further check the practicability and strength of the projected MCDM structures, we conducted a proportional investigation by utilizing other well-known methods to Example 7.2.1. By using the DM method given in (Zhang et al., 2017), we attain the RO as $H_4 \succ H_1 \succ H_2 \succ H_3$, and the best alternative remains same: H_4. On utilizing the possibility degree method provided by Garg and Kumar (2018b), we obtain the RO $H_4 \succ H_3 \succ H_2 \succ H_1$, and H_4 is the best alternative. Meanwhile, various researchers proposed the different AOs in the LIFNs context. Chen et al. (2015) defined the LIFWA AOs for aggregating the LIFNs. Therefore, by applying the LIFWA AO to aggregate the alternatives, information, linguistic score, and accuracy function to rank the alternatives, we get the RO $H_4 \succ H_3 \succ H_2 \succ H_1$. The superlative alternative H_4 is still the same as best alternative obtained by the proposed Approach I. On utilizing the power AOs as provide by Liu and Qin (2017), the finest option remains the same as that proposed Approach I. More detailed results of the comparative study are specified in Table 7.1. Hence, we obtain the same optimal alternative in the comparative study. However, the proposed MCDM method is more

TABLE 7.1
Results of the Comparative Study

Existing Methods	Performance of the Alternatives				Ranking
	H_1	H_2	H_3	H_4	
Zhang et al. (2017)	0.5339	0.4020	0.4266	0.6554	$H_4 \succ H_1 \succ H_3 \succ H_2$
Liu and Qin (2017)	2.9792	3.1431	3.6216	3.7452	$H_4 \succ H_3 \succ H_2 \succ H_1$
Chen et al. (2015)	2.9762	3.1826	3.6170	3.6762	$H_4 \succ H_3 \succ H_2 \succ H_1$
Zhang (2014)	5.4881	5.5913	5.8085	5.8381	$H_4 \succ H_3 \succ H_2 \succ H_1$
Garg and Kumar (2018b)	0.1909	0.2547	0.2453	0.3091	$H_4 \succ H_2 \succ H_3 \succ H_1$
Proposed method	2.1488	2.5025	2.7713	3.9587	$H_4 \succ H_3 \succ H_2 \succ H_1$

applicable and reliable because AOs bear a significant amount of flaws in the aggregating process, that if taking one value $\langle [s_h, s_h], [s_0, s_0] \rangle$ or $\langle [s_0, s_0], [s_h, s_h] \rangle$ out of the n input values for any alternative then achieving the aggregating value remains same $\langle [s_h, s_h], [s_0, s_0] \rangle$ or $\langle [s_0, s_0], [s_h, s_h] \rangle$ respectively which is unreasonable.

7.5 CONCLUSION

LIFSs are the most powerful tool in expressing the qualitative uncertainty of the MCDM process. Therefore, the entire objective of this chapter is about the LIFVs environment for solving the MCDM issues. In order to execute this, the proponents have constructed an entropy measure (EM) for measuring the fuzziness of the LIFS, which will be helpful for the DMks in terms of measuring uncertainty and obtaining the criterion weight of the MCDM issues. The criterion's weight has a fundamental role throughout the whole MCDM process, and it directly affects the ranking of alternatives of MCDM issues. Therefore, based on the proposed EM, two MCDM methods for solving the MCDM issues under the LIFVs context are utilized to handle realistic cases to exhibit the efficiency of the developed MCDM approaches. It has been inferred from the calculated outcomes that the proposed MCDM methods are sensible, feasible, and able to provide an advantageous way in dealing with MCDM problems in the context of LIFV. The proposed MCDM approaches overcome the drawbacks of the AOs and easily handle critical MCDM issues. In the future, the researchers shall develop the proposed methods for other environments, such as spherical fuzzy sets, linguistic interval-valued intuitionistic fuzzy sets, and complex IFS. Furthermore, the authors will develop new applications in other fields, such as reliability analysis, engineering, military, medical diagnosis, and pattern recognition, among others.

REFERENCES

Atanassov, K., & Gargov, G. (1989). Interval valued intuitionistic fuzzy sets. *Fuzzy Sets and Systems*, *31*(3), 343–349.
Atanassov, K. T. (1986). Intuitionistic fuzzy sets. *Fuzzy Sets and Systems*, *20*(1), 87–96.

Chen, Z., Liu, P., & Pei, Z. (2015). An approach to multiple attribute group decision making based on linguistic intuitionistic fuzzy numbers. *International Journal of Computational Intelligence Systems*, *8*(4), 747–760.

Deluca, A., & Termini, S. (1993). A definition of a non-probabilistic entropy in the setting of fuzzy sets theory. In *Readings in Fuzzy Sets for Intelligent Systems* (pp. 197–202). Morgan Kaufmann. DOI: https://doi.org/10.1016/B978-1-4832-1450-4.50020-1

Deluca, A., & Termini, S. (1972). A definition of a non probabilistic entropy in the setting of fuzzy set theory. *Information and Control*, *20*(4), 301–312.

Deschrijver, G., & Kerre, E.E. (2002). A generalization of operators on intuitionistic fuzzy sets using triangular norms and conorms. *Fuzziness and Uncertainty Modelling*, *8*(1), 19–27.

Dong, Y., Xu, Y., Li, H., & Feng, B. (2010). The OWA-based consensus operator under linguistic representation models using position indexes. *European Journal of Operational Research*, *203*(2), 455–463.

Garg, H. (2016a). Generalized intuitionistic fuzzy interactive geometric interaction operators using Einstein t-norm and t-conorm and their application to decision making. *Computers & Industrial Engineering*, *101*, 53–69.

Garg, H. (2016b). Some series of intuitionistic fuzzy interactive averaging aggregation operators. *SpringerPlus*, *5*(1), 1–27.

Garg, H., & Kaur, J. (2018). A novel (R, S)-norm entropy measure of intuitionistic fuzzy sets and its applications in multi-attribute decision-making. *Mathematics*, *6*(6), 92.

Garg, H., & Kumar, K. (2017). A novel correlation coefficient of intuitionistic fuzzy sets based on the connection number of set pair analysis and its application. *Scientia Iranica*, *25*(4), 2373–2388.

Garg, H., & Kumar, K. (2018a). Group decision making approach based on possibility degree measures and the linguistic intuitionistic fuzzy aggregation operators using Einstein norm operations. *Journal of Multiple-Valued Logic & Soft Computing*, *31*, 175–209.

Garg, H., & Kumar, K. (2018b). Some aggregation operators for linguistic intuitionistic fuzzy set and its application to group decision-making process using the set pair analysis. *Arabian Journal for Science and Engineering*, *43*(6), 3213–3227.

Garg, H., & Kumar, K. (2019). Linguistic interval-valued Atanassov intuitionistic fuzzy sets and their applications to group decision-making problems. *IEEE Transactions on Fuzzy Systems*, *27*(12), 2302–2311.

Garg, H., & Kumar, K. (2020). A novel exponential distance and its based TOPSIS method for interval-valued intuitionistic fuzzy sets using connection number of SPA theory. *Artificial Intelligence Review*, *53*(1), 595–624.

Gupta, P., Mehlawat, M.K., Grover, N., & Pedrycz, W. (2018). Multi-attribute group decision making based on extended {TOPSIS} method under interval-valued intuitionistic fuzzy environment. *Applied Soft Computing*, *69*, 554–567.

Herrera, F., & Martinez, L. (2001). A model based on linguistic 2-tuples for dealing with multigranular hierarchical linguistic contexts in multi-expert decision-making. *IEEE Transactions on Systems, Man, and Cybernetics, Part B (Cybernetics)*, *31*(2), 227–234.

Jiang, W., & Wei, B. (2018). Intuitionistic fuzzy evidential power aggregation operator and its application in multiple criteria decision-making. *International Journal of Systems Science*, *49*(3), 582–594.

Kumar, K., & Garg, H. (2018). {TOPSIS} method based on the connection number of set pair analysis under interval-valued intuitionistic fuzzy set environment. *Computational and Applied Mathematics*, *37*(2), 1319–1329.

Kumar, K., Dhillon, V.S., Singh, P. L., & Sindhwani, R. (2019). Modeling and analysis for barriers in healthcare services by ISM and MICMAC analysis. In *Advances in Interdisciplinary Engineering* (pp. 501–510). Springer, Singapore. https://doi.org/10.1007/978-981-13-6577-5_47

Liu, J. C., & Li, D. F. (2018). Corrections to "{ TOPSIS}-based nonlinear-programming methodology for multi-attribute decision making with interval-valued intuitionistic fuzzy sets" [Apr 10 299-311]. *IEEE Transactions on Fuzzy Systems*, *26*(1), 391.

Liu, M., & Ren, H. (2014). A new intuitionistic fuzzy entropy and application in multi-attribute decision making. *Information*, *5*(4), 587–601.

Liu, P., & Liu, X. (2017). Multiattribute group decision making methods based on linguistic intuitionistic fuzzy power Bonferroni mean operators. *Complexity*, *2017*, Article ID 3571459, 15 pages.

Liu, P., & Qin, X. (2017). Power average operators of linguistic intuitionistic fuzzy numbers and their application to multiple-attribute decision making. *Journal of Intelligent & Fuzzy Systems*, *32*(1), 1029–1043.

Liu, P., & Wang, P. (2017). Some improved linguistic intuitionistic fuzzy aggregation operators and their applications to multiple-attribute decision making. *International Journal of Information Technology & Decision Making*, *16*(03), 817–850.

Mittal, V.K., Sindhwani, R., Shekhar, H., & Singh, P.L. (2019). Fuzzy AHP model for challenges to thermal power plant establishment in India. *International Journal of Operational Research*, *34*(4), 562–581.

Peng, H., Wang, J., & Cheng, P. (2018). A linguistic intuitionistic multi-criteria decision-making method based on the Frank Heronian mean operator and its application in evaluating coal mine safety. *International Journal of Machine Learning and Cybernetics*, *9*, 1053–1068. https://doi.org/10.1007/s13042-016-0630-z

Shannon, C.E. (1948). A mathematical theory of communication. *The Bell System Technical Journal*, *27*(3), 379–423.

Sindhwani, R., Mittal, V.K., Singh, P.L., Kalsariya, V., & Salroo, F. (2018). Modelling and analysis of energy efficiency drivers by fuzzy ISM and fuzzy MICMAC approach. *International Journal of Productivity and Quality Management*, *25*(2), 225–244.

Szmidt, E., & Kacprzyk, J. (2001). Entropy for intuitionistic fuzzy sets. *Fuzzy Sets and Systems*, *118*(3), 467–477.

Verma, R., & Sharma, B.D. (2013). Exponential entropy on intuitionistic fuzzy sets. *Kybernetika*, *49*(1), 114–127.

Wang, J.-Q., & Wang, P. (2012). Intuitionistic linguistic fuzzy multi-criteria decision-making method based on intuitionistic fuzzy entropy. *Control and Decision*, *27*(11), 1694–1698.

Wang, W., & Liu, X. (2011). Intuitionistic fuzzy geometric aggregation operators based on Einstein operations. *International Journal of Intelligent Systems*, *26*(11), 1049–1075.

Wang, W., & Liu, X. (2012). Intuitionistic fuzzy information aggregation using{ Einstein} operations. *IEEE Transactions on Fuzzy Systems*, *20*(5), 923–938.

Wei, C.P., Gao, Z. H., & Guo, T.-T. (2012). An intuitionistic fuzzy entropy measure based on trigonometric function. *Control and Decision*, *27*(4), 571–574.

Xian, S., Jing, N., Xue, W., & Chai, J. (2017). A New Intuitionistic Fuzzy Linguistic Hybrid Aggregation Operator and Its Application for Linguistic Group Decision Making. *International Journal of Intelligent Systems*, *32*(12), 1332–1352. https://doi.org/10.1002/int.21902

Xu, Y., Merigó, J.M., & Wang, H. (2012). Linguistic power aggregation operators and their application to multiple attribute group decision making. *Applied Mathematical Modelling*, *36*(11), 5427–5444. https://doi.org/10.1016/j.apm.2011.12.002

Xu, Z., & Chen, J. (2007). On geometric aggregation over interval-valued intuitionistic fuzzy information. *Fourth International Conference on Fuzzy Systems and Knowledge Discovery (FSKD 2007)*, *2*, 466–471.

Xu, Zeshui. (2004). A method based on linguistic aggregation operators for group decision making with linguistic preference relations. *Information Sciences*, *166*(1), 19–30.

Xu, Zeshui. (2007a). Intuitionistic fuzzy aggregation operators. *IEEE Transactions on Fuzzy Systems*, *15*(6), 1179–1187.

Xu, Zeshui. (2007b). Methods for aggregating interval-valued intuitionistic fuzzy information and their application to decision making [J]. *Control and Decision*, *2*, 19.

Xu, Zeshui, & Yager, R.R. (2006). Some geometric aggregation operators based on intuitionistic fuzzy sets. *International Journal of General Systems*, *35*(4), 417–433.

Xu, Zeshui, & Zhao, N. (2016). Information fusion for intuitionistic fuzzy decision making: an overview. *Information Fusion*, *28*, 10–23.

Zadeh, L.A. (1965). Fuzzy sets. *Information and Control*, *8*(3), 338–353.

Zadeh, L.A. (1975). The concept of a linguistic variable and its application to approximate reasoning - I. *Information Sciences*, *8*(3), 199–249.

Zhang, Hongyu, Peng, H., Wang, J., & Wang, J. (2017). An extended outranking approach for multi-criteria decision-making problems with linguistic intuitionistic fuzzy numbers. *Applied Soft Computing*,*59*, 462–474.

Zhang, Huimin. (2014). Linguistic intuitionistic fuzzy sets and application in { MAGDM}. *Journal of Applied Mathematics*, *2014*, Article ID 432092, 11 pages. https://doi.org/10.1155/2014/432092

Zhang, Q.S., & Jiang, S.Y. (2008). A note on information entropy measures for vague sets and its applications. *Information Sciences*, *178*(21), 4184–4191.

8 An Insight into Decision-Making Using the Fuzzy Analytic Hierarchy Process (FAHP)

Nikhil Dev and Rajeev Saha

CONTENTS

8.1 Introduction 139
8.2 Complexity of Analysis 140
8.3 Data Collection and Fuzzy Logic 144
8.4 Decision-Making with the Analytic Hierarchy Process 151
8.5 Conclusion 153
References 153

8.1 INTRODUCTION

Fuzzy means lack of clarity. Engineering project management includes a variety of information including time, cost, quality analysis, resource management, procurement of material, resource utilization efficiency, safety, environment, stakeholder management, *etc*. The proportion of contribution from these attributes in decision-making varies from place to place and time to time. Variability leads to complexity in decision-making. It is required to obtain appropriate data for system design development and analysis. In the modern world of competition, most systems are dynamic in nature. A dynamic system is a system that is capable of improving itself constantly (Saaty, 1980). Due to a dynamic system, data availability for decision-making is not available in appropriate proportion and is rather concisely condensed. Variable data arrangement in terms of order and its interpretation is one of the complex tasks performed by decision-makers. This is similar to the Heisenberg uncertainty principle, which states that things in this world are quantized and contains a minimum amount of uncertainty. Uncertainty and fuzziness are two different sides of the same coin. Fuzziness in the system is due to a lack of clarity in data, and uncertainty is because of the non-availability of measuring instruments. However, both of the methodologies are capable of guiding human beings in decision-making with the use of mathematical modeling. Conclusively, errors cannot be avoided completely in any system.

FIGURE 8.1 Data available is fuzzy in nature, and it has to be inferred.

The process is complex due to lengthy mathematical equations. These equations are the combination of calculus and other parts of mathematics. Therefore, it is not easy to solve the problem without the use of computational techniques. Nonetheless, it is required to filter the data with the help of fuzzy logic, as illustrated in Figure 8.1. Afterward, it will be processed with the help of AHP and certain suitable computational tools. This process includes three steps 1) Fuzzification 2) Fuzzy Inference, and 3) Defuzzification (Zadeh, 1971). The data may be available with the user, or it may be collected with the help of a survey. A fundamental aspect of process planning is developing a system model that defines the functional relationship between system inputs and outputs. The rapport of input and output imitates the objectives of manufacturing or system development. However, the process of planning and analysis is a complex process.

8.2 COMPLEXITY OF ANALYSIS

The thought process for a human being is generally not very common (Saaty & Vargas, 2006). This means that in society, only a few people are capable of a very particular kind of thought. As the population of the world is significantly large, therefore, thoughts are also rather variable. At this point of variable thoughts, it is important to understand that scattered data for any particular analysis is obtained. Data is not quite easily clear that any quick decision is possible. The same kind of behaviour is observed at the level of machines. No two similar machines behave in exactly the same manner. Therefore, in this world, it is not suitable to think that an exact and systematic decision, which may act as a silver bullet, is possible.

Therefore, in the process of analysis, three-step methodologies are followed:

1. Data collection and analysis.
2. Application of certain decision-making techniques.
3. Use of computational tools.

Data available in literature or from surveys are not always perfect. Moreover, data may be rather large, and the selection of certain suitable decision-making techniques is paramount. This data will be processed with a suitable computer programming tool, as explained in Figure 8.2. The information available to us from any type of source can be qualitative or quantitative. Sometimes, information is also imperfect, as illustrated in Figure 8.3. All of these types of information are to be processed so that data can be streamlined and analyzed easily. For decision-making, FAHP is an MADM technique. The data processed is used for making the final decision with the help of FAHP. Computational tools are selected, as per the

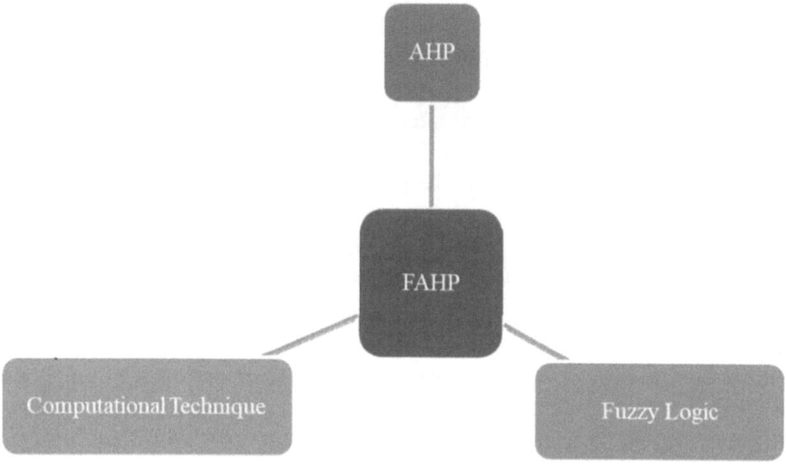

FIGURE 8.2 FAHP solution methodology is the combination of (a) Fuzzy Logic, (b) AHP, and (c) Computational Technique.

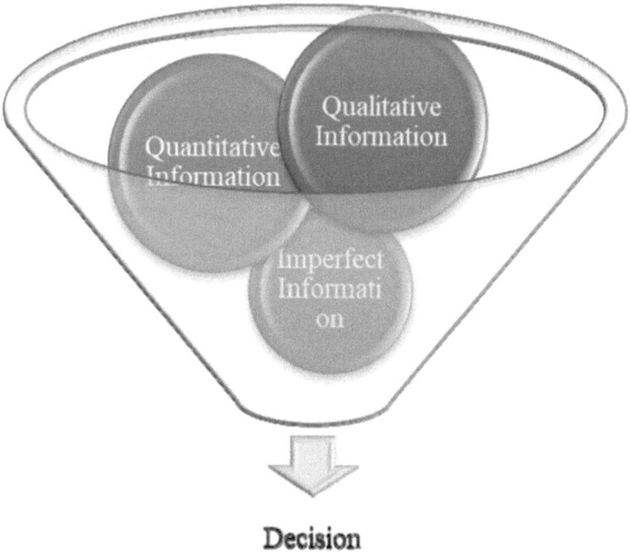

FIGURE 8.3 In sizeable organizations, the information available is (a) Qualitative, (b) Quantitative, and (c) Imperfect.

requirement of the system. This is an enormous subject in itself and cannot be discussed in this chapter in detail. However, from the literature survey, it has been discovered that a lot of research on using AHP and FAHP is in progress. Aghdaie et al. (2013) carried out a market segment evaluation and selection while using FAHP and COPRAS-G. FAHP can be used for supplier selection in different market types (Alinezad et al., 2013; Chan et al., 2008; Chan et al., 2013). Resource management is a fairly complex process, and it can be simplified with the help of FAHP (Aryafar et al., 2013; Anagnostopoulos & Petalas (2011); Azarnivand et al., 2015). Other areas of quality improvement, such as maintenance (Azadeh & Zadeh, 2016) and bench-marking (Carnero, 2014; Hosseini Firouz & Ghadim, 2016; Turskis, 2012) are achieved with the help of AHP.

Every decision-maker is sitting in a jostling ship with a cup of tea in his hand. His ability to sip tea will depend on skill and practice. However, this may also be proven wrong in some cases. Nonetheless, even if the decision-making is right, it is still fuzzy in nature. Even with an inappropriate (but not wrong) decision, there is always the availability of a second opportunity, which may be less beneficial. At least at this point, it can be said that it is a net that we are dealing with instead of a straight chain.

The meaning of fuzzy is 'not very clear,' and in reality, most things are fuzzy in nature. Therefore, we need to clarify or acquire certain tools which can turn something vague into clear terms, as further portrayed in Figure 8.4. In mathematics, fuzzy logic is used to handle concepts where the data available is rather unclear. The concept of fuzzy logic has been used by ancient mathematicians. With the help of fuzzy logic, qualitative information can be converted into quantitative information.

Fuzzy relies on the observation that many people, or almost everyone, make decisions based on experience and non-numerical information. This information is obscure. Therefore, in order to deal with imprecise information, there is a need for certain mathematical tools that can convert qualitative information into useful information. Furthermore, with the help of a mathematical tool, we can convert this vagueness into some useful terms. Things are fuzzy in nature due to the complex thought process of human beings. An elaborate neural network of the mind operating at a very low voltage is capable of processing large amounts of information.

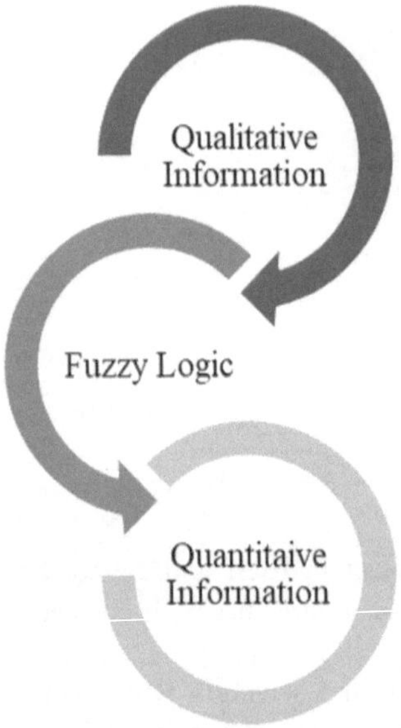

FIGURE 8.4 Fuzzy logic has the power to convert qualitative information into quantitative information.

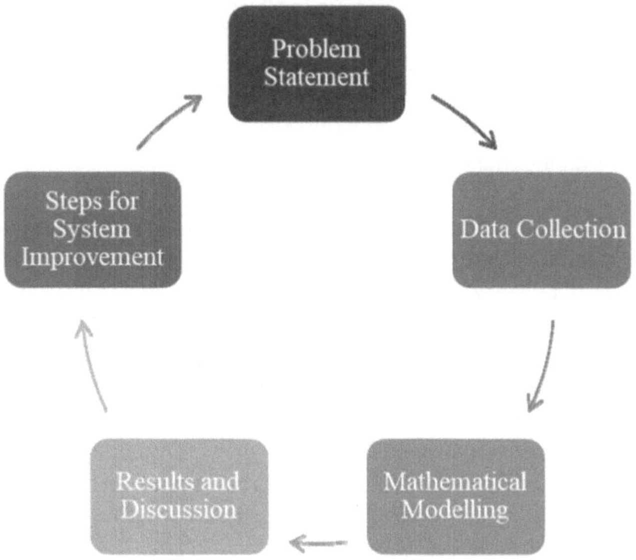

FIGURE 8.5 Road Map to Improving any Organizational Structure.

The calculation speed is low due to the semisolid nature of the mind, and electrons travelling speed is reduced in comparison to a solid-state system.

Decision-making with multiple attributes is a process that is used by scientists and engineers for many different kinds of real-time problems. Decision-making is a technique that will consider many different parametres, and these parametres will also be interdependent. At this point, it is worth mentioning that improvement in any organization is a cyclic process, as presented in Figure 8.5. This process begins with a certain problem statement. Next, the data related to the problem has to be collected, along with parametres related to this problem. The selection of parametres and the choice of interdependence are completely based upon the knowledge of the person dealing with these parametres. If the person involved in data collection is highly knowledgeable, then the results will be more accurate. At this point, in terms of accuracy, the person dealing with the parametre will be collecting data from another person, and this data collection will be completely based upon his or her experience. If a person is more experienced in his work, then the methodology will be more appropriate. The process of decision-making can be summarized as follows (Saaty & Vargas, 2000):

1. Define the problem and determine the parametres affecting the solution of the problem.
2. Select the type of methodology which is most appropriate for decision-making.
3. Develop a system structure, as per the availability of mathematical tools.
4. Use the results obtained from the system structure for organizational development and betterment.

8.3 DATA COLLECTION AND FUZZY LOGIC

Data collection falls under two broad categories:

1. Primary data collection.
2. Secondary data collection.

Primary data collection is the process of gathering raw and original data by a researcher for a specific research purpose. Data is classified into two types: qualitative and quantitative. Secondary data is the data collected by an individual who is not the original user. The importance of data can be estimated from Figure 8.6.

At this point, it can be said that both data collection and data interpretation are dependent upon the expertise of the concerned person. If a person is adept in data collection and interpretation, then the accuracy of the results will also be very high. This process also signifies that people are dealing with qualitative terms or parametres instead of with quantitative terms or parametres at all times in reality. In decision-making applications, fuzzy logic is used to convert qualitative data into quantitative data, as observed in Figure 8.7.

The fuzzy number system consists of the types: triangular fuzzy number, triangular intuitionistic fuzzy number, interval-valued triangular fuzzy number, trapezoidal fuzzy number, intuitionistic trapezoidal fuzzy number, internal-valued trapezoidal fuzzy number, *etc*. Fuzzy logic can be applied with many MADM techniques, such as ordered weighted aggregation operators, weighted geometric

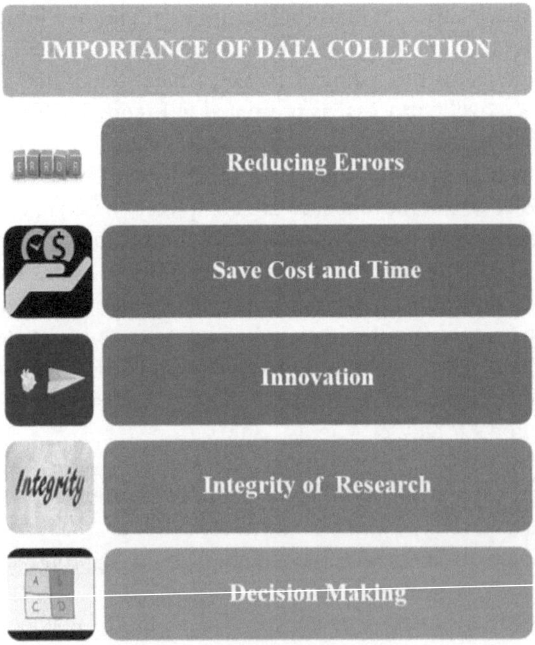

FIGURE 8.6 Importance of Data in Decision-Making.

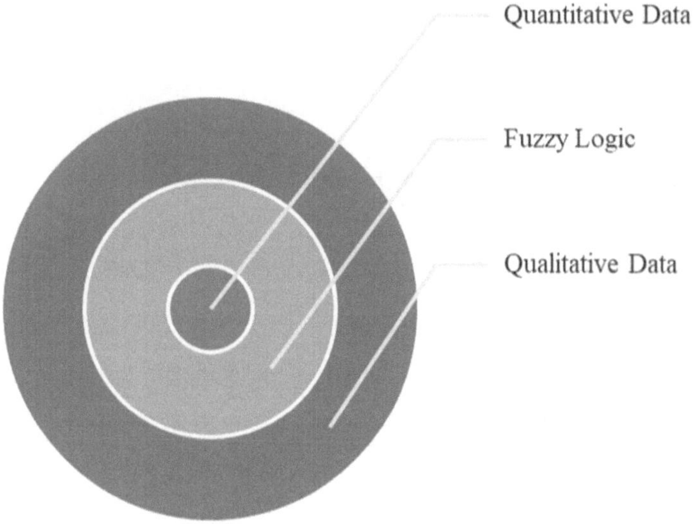

FIGURE 8.7 Qualitative data can also be converted into quantitative data with the help of fuzzy logic.

aggregation operators, TOPSIS, the analytic hierarchy process method, the grey relational analysis method, similarity measures, Graph-Theoretic Analysis. COPRAS, *etc*. Fuzzy numbers were introduced by Zadeh (1965).

Fuzzy analysis is useful in developing more accurate control systems. The applications of fuzzy control in the modern control system design show that FLCs yield results that are superior to those obtained by conventional control algorithms. The objective of using fuzzy logic with MADM techniques is to convert the qualitative and diversified data into certain useful quantitative data. At present, fuzzy logic can be implemented in real-life operating systems with different sizes and capabilities, ranging from small micro-controllers to large, networked, workstation-based control systems. It is implemented with the help of both hardware and software.

Everything in this world is fuzzy in nature. Even two components made by the same machine at the same time are not exactly identical. Data collected from different sources cannot be identical. In this case, data condensation and removal of extreme data are necessary. This can be observed in Figure 8.8, where data is scattered. The data presented in Figure 8.8 is quantitative. In this figure, some data is on the extremely lower side, and some data is on the extremely upper side. These types of inferences (e.g. extremely low, extremely high) are qualitative in nature. Secondly, the usefulness of any data is user-application-dependent.

Qualitative data can be of two types: (a) Nominal Data and (b) ordinal data. The different types of data are represented in Figure 8.9. The nominal scale is used for labeling variables without any quantitative value (e.g black, white, *etc.*), while the ordinal scale measures non-numeric concepts (e.g. satisfaction, happiness, discomfort, *etc*). Discrete data cannot be counted precisely. Rather, it involves real

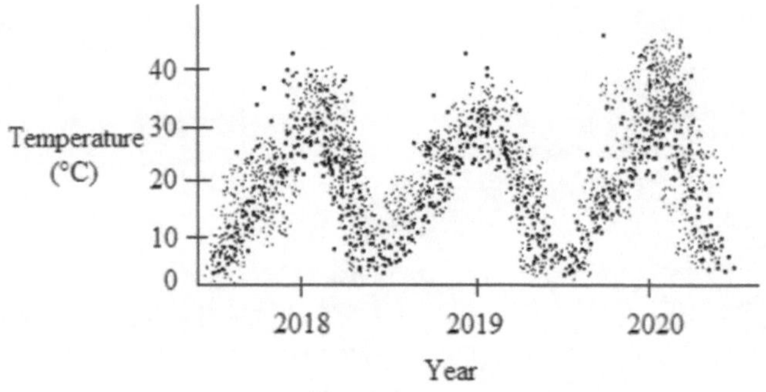

FIGURE 8.8 Temperature Distribution of a City for Three Years.

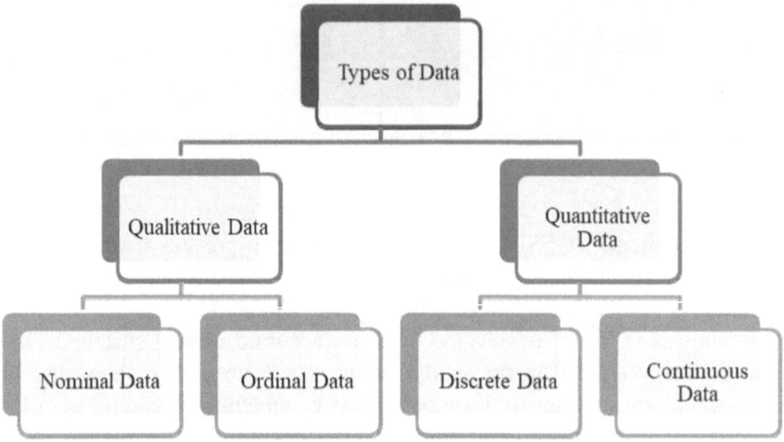

FIGURE 8.9 Data and Its Different Types.

numbers and integers. For instance, the number of eggs you bought will be 1, 2, *etc.* and not 1.5, 2/5. Some of the data is continuous in nature, such as the length of a bar that can be represented at progressively more precise measurements scales – metres, centimetres, millimetres, and beyond – so length is a type of continuous data.

Data collection and assessment are also required for risk reduction also, as presented in Figure 8.10. A fully automated data transfer and processing system minimizes the need for manual manipulation and calculation of raw data. A Data Transfer Unit (DTU) is used to eliminate human error while processing the data. This type of process is helpful in the aviation industry, for example.

A large amount of data is moved in the Data Processing Unit (DPU), which requires considerable time and network bandwidth. Therefore, it is required to scale-up the workflow in order to process the data. Data processing is represented in Figure 8.11. A rather important challenge to overcome is processing the petabytes of data that are collected during the development and testing of autonomous driving systems.

FIGURE 8.10 Benefit of Data Collection and Assessment.

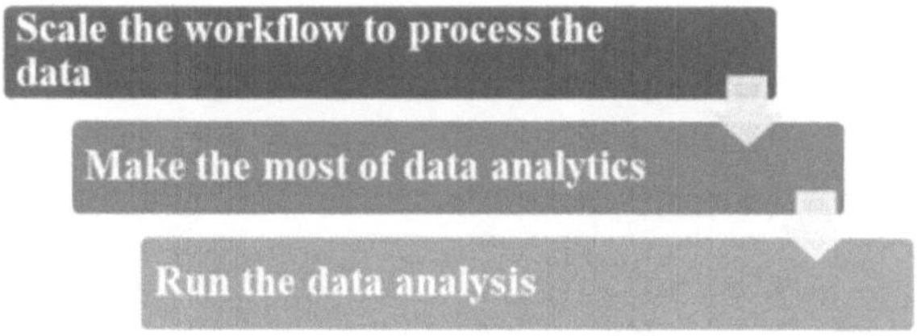

FIGURE 8.11 Systematic Flow Representing the Method of Data Processing.

Fuzzy logic is useful for commercial and practical purposes because of the following reasons:

- It can control machines and consumer products in a sufficient way.
- It may not be precisely cognitive, but its cognizance is acceptable.
- Fuzzy logic deals with uncertainty in a more creditable way.

Fuzzy logic can be understood with the help of an example. Suppose in some automobile industries, the quality of manufacturing for a particular component is average. Here, the qualitative term 'average' is related to the reliable working life of the component. The component life will be affected by the speed of travelling of the vehicle and stresses generated in the component due to the weight of the occupants. According to the conditions mentioned, the vehicle can bear a maximum load of 700 Kg, and its maximum speed can be 100 Km/hr. The optimum speed at which the vehicle proves the best fuel economy is 60 Km/hr with a load of 400 Kg. If the load is increased, then the operating speed of the vehicle has to be kept low in order to avoid the failure of the components. At lower speed, the time for the journey will be increased, and hence, transportation cost will also be increased. Therefore, at this point, the end conditions are clear, but the intermediate condition is not clear. In this case, some factors such as road condition, weather, fuel quality, *etc.* may also come into the picture during real-time operation. All of these factors can be analyzed one by one or all at the same time. In this direction, we are taking load and road condition which will decide the speed of the vehicle.

To develop fuzzy logic, create a matrix of load and road conditions. The matrix representation is the combination of inheritance and interdependence. Furthermore, in the matrix, all of the elements are interdependent. Interdependence can also be in tabular form, as represented in Table 8.1.

TABLE 8.1
Qualitative Analysis of Road Condition, Load, and Speed of the Vehicle

Load/Road Condition	Maximum	High	Medium	Low	Minimum
Excellent	D	OK	OK	I	I
Good	D	OK	OK	I	I
Normal	D	D	OK	I	I
Bad	D	D	D	OK	OK
Worst	Minimum Speed	D	D	D	OK

In the table above, the different symbols are defined as the following:

D = Decrease speed
OK = No change in speed
I = Increase Speed

In Table 8.1 qualitative analysis is carried out while considering the effect of load and road condition on the speed of the vehicle. This qualitative analysis is insufficient in developing a controller. The analysis presented in Table 8.1 is quite elementary. In real-time, the quantity of the data will be enormous. It has to be managed, as mentioned in Figure 8.11. For the design and development of the controller, it is required to construct a set of rules into the knowledge base structure comprising of IF-THEN-ELSE, as listed in Table 8.2.

There are many methods used to define the fuzzy sets. In literature, it is determined that fuzzy sets are often demarcated as triangle or trapezoid-shaped curves, as depicted in Figure 8.12. These can be used for the quantification of the qualitative data. In automotive vehicles, problem speed is defined only in qualitative terms. As such, it is mentioned only as 'increase speed' or 'decrease speed'. It is not a quantitative measurement. However, we can now have low speed, medium speed, or high speed. Supposing the vehicle is moving at low speed, and this was due to the bad condition of the rod. Now, for some part of the journey, road condition improved. The vehicle can then run at medium speed. Therefore, as we will proceed

TABLE 8.2
Condition Action Table w.r.t. Fuzzy Logic

Serial Number	Condition	Action
1	If load is maximum (near to 700 Kg), and road condition is normal or bad, then	Decrease speed
2	If road condition is good or excellent, and load medium then,	No change in speed
3	If road condition is normal, good, or excellent, and load is also low then,	Increase speed

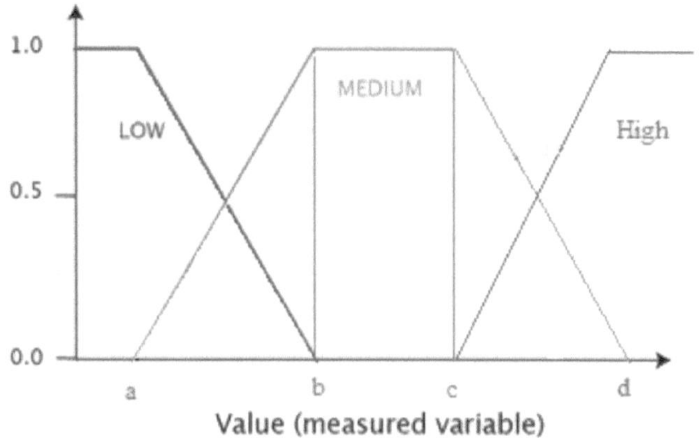

FIGURE 8.12 Trapezoid-Shaped Curve Defining the Fuzzy Set.

further on the trapezoidal curve, we are also progressing away from low speed and moving towards medium speed. Moreover, each value will have a slope where one value is decreasing and the other value is increasing. There will be a peak for each medium and low speed, where the value is equal to 1. The length of the peak point can be 0 or greater. Each curve will bear three parts: i) Decreasing Value, ii) Constant Value, ad iii) Increasing Value. The functions defining such curves are sigmoid functions. The membership value or degree of membership in fuzzy logic quantifies the grade of membership of any element in x to the fuzzy set A, where x is defined as μ:x → [0, 1]. The mathematically trapezoidal function, triangular function, and the Gaussian function as represented in Figure 8.12, Figure 8.13, and Figure 8.14 respectively.

$$\mu(x) = \begin{cases} 0, & (x < a)\ or\ (x > a) \\ \frac{x-a}{b-a}, & a \le x \le b \\ 1, & b \le x \le c \\ \frac{d-x}{d-c}, & c \le x \le d \end{cases} \tag{8.1}$$

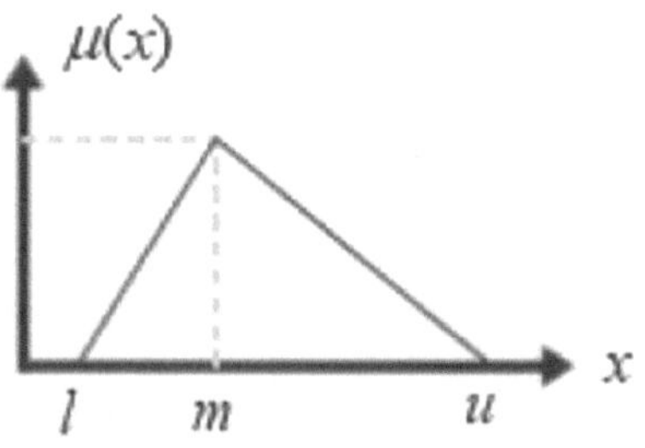

FIGURE 8.13 Triangular Membership Function.

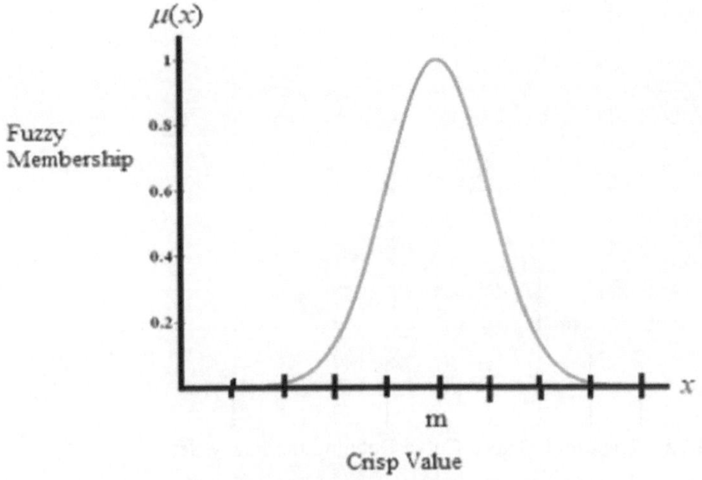

FIGURE 8.14 Gaussian Membership Function.

$$\mu(x) = \begin{cases} 0, & x \le l \\ \frac{x-l}{m-l}, & l < x \le m \\ \frac{u-x}{u-m}, & m < x \le u \\ 0, & x \ge b \end{cases} \tag{8.2}$$

$$\mu(x) = e^{-\frac{(x-m)^2}{2k^2}} \tag{8.3}$$

Where m is the central value with standard deviation **k > 0**. The smaller is the value of k, narrower is the shape of bell.

The mathematical expression used for the trapezoidal function, triangular function, and Gaussian function is represented by the expression (8.1), (8.2), and (8.3) respectively. The flow chart for applying fuzzy logic is as represented in Figure 8.15. Fuzzy logic has many applications in real life, as further explained in Figure 8.16.

Fuzzy logic controllers (FLCs) have the following advantages over conventional controllers:

- Simplicity and flexibility in system implementation.
- Ability to handle problems with imprecise and incomplete data.
- Modelling of nonlinear functions with random complexity.
- Clarity and pliability that models human reasoning.
- Capability to deal with imprecise and incomplete data.
- Use of linguistic variables.

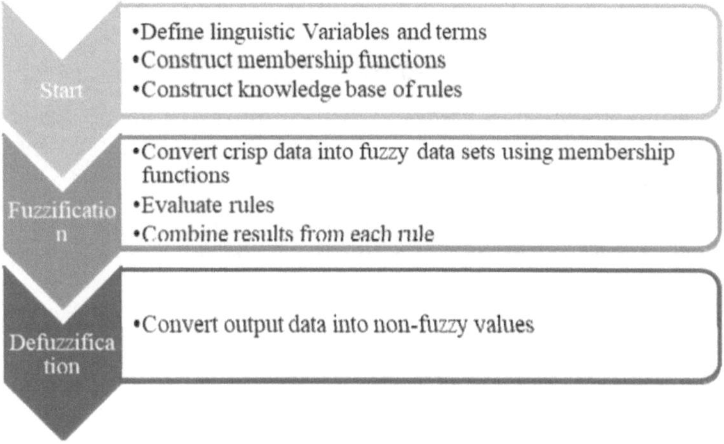

FIGURE 8.15 Flow Chart of Fuzzy Logic.

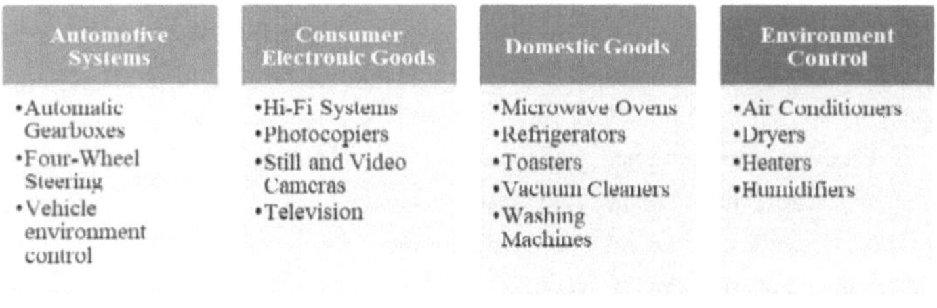

FIGURE 8.16 Use of Fuzzy Logic in Various Applications.

Due to these advantages, fuzzy logic is gaining more and more applications in real life, as exemplified in Figure 8.16.

8.4 DECISION-MAKING WITH THE ANALYTIC HIERARCHY PROCESS

The Analytic Hierarchy Process (AHP) is a theoretical analysis of decision-making problems (Saaty & Alexander, 1989). It is based upon the pairwise comparison. In the present work, we have considered an example of an automotive vehicle. The problem with the automotive vehicle is that if the road condition is bad, then speed must be kept low. Therefore, one factor depends on the other. On the other hand, if the speed of the vehicle is already low, then road condition is not a very important factor. This type of comparison is called a pairwise comparison. For this concept, the comparison of qualitative terms is converted into quantitative terms, as accomplished in Table 8.3.

With the conversion of qualitative terms into quantitative terms, it becomes easy to quantify the pairwise comparison matrix. In this matrix, a diagonal element represents the relation of the element with itself. It is always considered as one. The value for the non-diagonal elements is taken on a scale of one to nine, and this is represented in Table 8.4.

TABLE 8.3
Quantification of Factors (Saaty, 2005)

S. No.	Qualitative Measure of Parametres	Fuzzy Notion	Assigned Value of Parametres S_i
1	The effect of one parametre on the other is exceptionally low.	Exceptionally low	1
2	The effect of one parametre on the other is very low, and with a change in one parametre, the other is not much affected.	Very low	2
3	This value will be assigned when the effect of one parametre on the other is low. This effect should be higher than the previous category.	Low	3
4	With this quantification, the inheritance value is changed from low to below average. It is to represent that the impact of the parametre in the system is increasing.	Below average	4
5	It is one of the three broad categories, i.e. high, average, and low. The parametres falling in this category have a medium effect on the system.	Average	5
6	If it is discovered that parametre inheritance is somewhat more than average, then these are kept in this category. It is the category representing inheritance as higher than average.	More than average	6
7	In this category, parametres having high impact on the system are kept, and from this point, inheritance starts playing a role, which is having the slope higher than previous categories.	High	7
8	If the inheritance of any parametre category is very high, then it is covered in this section. This value should be somewhat less than isentropic.	Very high	8
9	This quantification represents that the parametres category has a very high impact on the system. This may be called exceptionally high. It may be an isentropic case.	Exceptionally high	9

TABLE 8.4
Pairwise Comparison Matrix

	Load	Road Condition	Speed
Load	1	3	1/7
Road condition	1/3	1	1/9
Speed	7	9	1

The methodology of AHP can be summarized as follows in the language of Saaty (1994):

1. Define the problem and objective function.
2. Structure the decision hierarchy.
3. Construct a set of pairwise comparison matrices.
4. Use the priorities obtained from the comparisons to weigh the priorities.

Therefore, it can be concluded at this point that quantification of parametres and their impact on one another is crucial to quantify. The fuzzy approach provides a better way to quantify parametres and their interdependence. The combination of Fuzzy Logic and AHP can be used to solve many different types of decision-making problems. In literature, FAHP has been used for many applications, such as: Product pricing, material selection, power plant selection, decision-making in the power plant sector, demand forecasting, system structure analysis, comparison of two or more similar systems, the health sector, sustainability analysis, *etc.*

Defuzzification is the procedure of representing a fuzzy set with a crisp number. There are many methods of defuzzification. The centre of area method (COA) is the most commonly used defuzzification method. It is also referred to as the centroid method. In this method, the centre of the area of a fuzzy set is determined, which helps in developing a crisp value. The centre of sums (COS) method and the mean of the maximum method are the other two alternative methods used for defuzzification. In the COS method of defuzzification, the overlapping area is counted twice. In order to select the appropriate method of analysis, expert opinion is required. Based on experience, sensitivity analysis can be carried out and the results obtained will be the most appropriate.

8.5 CONCLUSION

In literature, a number of techniques are available for decision-making. A suitable methodology is to be developed in order to utilize these techniques for different types of problems. In the present work, FAHP is described in a manner that any new researcher can gain a brief overview on the utilisation of FAHP for solving any modern, real-time problem. The details of the FAHP methodology will be described in the forthcoming section of the present work. At this point, it can be concluded that FAHP is a powerful tool in solving the problems in which the available data is fuzzy in nature.

REFERENCES

Aghdaie, M.H., Zolfani, S.H., & Zavadskas, E.K. (2013). Market segment evaluation and selection based on application of fuzzy AHP and COPRAS-G methods. *Journal of Business Economics and Management*, *14*(1), 213–233.

Alinezad, A., Seif, A., & Esfandiari, N. (2013). Supplier evaluation and selection with QFD and FAHP in a pharmaceutical company. *The International Journal of Advanced Manufacturing Technology*, *68*, 355–364.

Anagnostopoulos, K.P., & Petalas, C. (2011). A fuzzy multicriteria benefit–cost approach for irrigation projects evaluation. *Agricultural Water Management*, *98*(9), 1409–1416.

Aryafar, A., Yousefi, S., & Ardejani, F.D. (2013). The weight of interaction of mining activities: groundwater in environmental impact assessment using fuzzy analytical hierarchy process (FAHP). *Environmental Earth Sciences*, *68*(8), 2313–2324.

Azadeh, A. & Zadeh, S.A. (2016). An integrated fuzzy analytic hierarchy process and fuzzy multiple-criteria decision-making simulation approach for maintenance policy selection. *Simulation*, *92*(1), 3–18.

Azarnivand, A., Hashemi-Madani, F.S., & Banihabib, M.E. (2015). Extended fuzzy analytic hierarchy process approach in water and environmental management. *Environmental Earth Sciences*, *73*(1), 13–26.

Carnero, M.C. (2014). Multicriteria model for maintenance benchmarking. *Journal of Manufacturing Systems*, *33*(2), 303–321.

Cebeci, U. (2009). Fuzzy AHP-based decision support system for selecting ERP systems in textile industry by using balanced scorecard. *Expert Systems with Applications*, *36*(5), 8900–8909.

Chan, F.T.S., Kumar, N., Tiwari, M.K., Lau, H.C.W., & Choy, K.L. (2008). Global supplier selection: A fuzzy-ahp approach. *International Journal of Production Research*, *46*(14), 3825–3857.

Chan, H.K., Wang, X., White, G.R.T., & Yip, N. (2013). An extended fuzzy-AHP approach for the evaluation of green product designs. *IEEE Transactions on Engineering Management*, *60*(2), 327–339.

Hosseini Firouz, M., & Ghadimi, N. (2016). Optimal preventive maintenance policy for electric power distribution systems based on the fuzzy ahp methods. *Complexity*, 1–19.

Saaty, T.L. (1980). *The analytic hierarchy process*. McGraw-Hill, New York. International, Revised editions, Paperback *(1996, 2000)*, Pittsburgh: RWS Publications.

Saaty, T.L. (1994). How to make a decision: The analytic hierarchy process. *Interfaces*, *24*(6), 19–43.

Saaty, T.L. (2005). *Theory and applications of the Analytic Network Process*. RWS Publications, Pittsburgh, PA.

Saaty, T.L. & Alexander, J. (1989). *Conflict resolution: The analytic hierarchy process*. Praeger, New York.

Saaty, T.L. & Vargas, L.G. (2000). *Models, methods, concepts and applications of the analytic hierarchy process*. Kluwer Academic Publishers, Boston.

Saaty, T.L. & Vargas, L.G. (2006). *Decision making with the analytic network process: Economic, political, social and technological applications with benefits, opportunities, costs and risks*. Springer, New York.

Turskis, Z. (2012). Maintenance strategy selection using AHP and copras under fuzzy environment. *International Journal of Strategic Property Management*, *16*(1), 85–104.

Zadeh, L. (1965). Fuzzy sets. *Information and Control*, *8*, 338–353.

Zadeh, L.A. (1971). Quantitative fuzzy semantics. *Information Sciences*, *3*(2), 159–176.

9 Application of the Fuzzy Analytic Hierarchy Process for Solar Panel Vendor Selection

Rajeev Saha and Nikhil Dev

CONTENTS

9.1 Introduction 155
9.2 Evaluation Model for the Solar Panel Module 156
9.3 The Hierarchical Structure 157
9.4 Determination of Criteria and Sub-Criteria Weights 158
9.5 Critical Weight of Manufacturers, Experts, and Users 169
9.6 Evaluator's Estimation of the Performance of the Alternatives 169
9.7 Ranking of the Alternatives 173
9.8 Conclusion 175
References 176

9.1 INTRODUCTION

Technological advancements have given rise to increased demand for energy. As the world is moving from the internet age of Industry 3.0 to a connected and collaborative world based on the Internet of Things in Industry 4.0, the demand for energy has increased tremendously. Industry 4.0 envisages everything to be connected, wherein data needs to be stored and analyzed for a smart decision-making system to be in place for connected things. As everything is linked, a huge amount of data needs to be stored; hence, the requirement for energy will grow exponentially. This huge demand for energy can only be met through renewable sources. Solar energy is one such source and is in abundance and needs to be tapped smartly for better utilisation. The solar panel modules used for acquiring solar energy are continuously evolving to bear better efficiency.

Selecting an efficient solar energy system requires choosing the solar panel module vendor, and this can be done by using the Fuzzy Analytic Hierarchy Process (FAHP) under Multi-Criteria Decision-Making (MCDM). Decision-making using FAHP is preferred mainly due to AHP's inherent capability of handling intangibles and the ease with which the criteria can be structured and compared. Moreover, fuzzy theory helps in handling uncertainties in pairwise comparisons.

9.2 EVALUATION MODEL FOR THE SOLAR PANEL MODULE

A solar panel module that is worth consideration should be able to qualify for the minimum criteria set for it. Besides the price of solar panel modules, there are various other criteria necessary to produce a maximum return on investment. The solar module is the main constituent of a solar panel. These are arranged in an array to generate and supply electricity. Generally, each module is capable of generating 100 to 365 watts of power under standard test conditions (Wang & Tsai, 2018). The criteria worth acknowledging for the evaluation of the solar panel module are discussed below:

- The Module Efficiency is the percentage of sunlight that hits the solar panel and is converted into usable electricity. To obtain higher efficiency, a fewer number of panels should be utilized to constitute the solar panel system and meet the energy requirements (What Does Module Efficiency Mean?, December 2016).
- Power Sorting is another criterion worth recognizing as it helps in overcoming the loss of power due to a mismatch in the maximum power output of the connected modules. It has been observed that, in certain cases, the loss can be more than 10% (Webber & Riley, 2013).
- Lifetime Energy Yield is the amount of energy produced by the module in its lifetime. Nowadays, the life of the energy module being produced is around 25 to 30 years. Although generally, it has been discerned that each year, the energy yield of a module reduces by 1%. A significant amount of research shows that less than 1% loss is achieved (Jordan & Kurtz, 2013; Jordan et al., 2016).
- Under Standard Test Conditions (STC), the power generated is checked by keeping the solar cell and ambient temperature at 25 °C. This is far from reality, as power significantly drops at increased temperature. Under PVUSA Test Conditions (PTC), the power generated is confirmed by keeping solar cell temperature at 45 °C and ambient temperature at 20 °C, along with a cooling wind speed of 1 m/s. The power rating thus provided is more realistic.
- The warranty of solar modules is iteratively important, as it safeguards against the efficiency and the time period up to which it is expected to work accurately.
- With better government support, more and more entities will install solar systems for meeting energy requirements. However, at the same time, the price of the solar system could be a deterring and limiting factor.
- The cost per watt of power generated by solar panels provides a real return on investment.
- The dimension of the solar panel provides the details of the area being covered for the generation of the required amount of power.
- The weight of the solar panels provides the data of how many solar panels can

be installed on a rooftop. In such a scenario, this limits the amount of the power installation capacity, as it is directly related to the power generated.

- The durability of the solar modules is again important, as the replacement has to be done thereafter. Usually, solar panels are durable for a period of 25 years.
- Corrosion Resistance of the solar panel helps in increasing the durability and hence, the life of the product.
- The Fill Factor of a solar cell is a key element in evaluating the performance of the solar system. Fill Factor represents the ratio of the maximum obtainable power to the product of the open-circuit voltage and short-circuit current.
- The cumulative global PV waste in 2016 was 250,000 metric tonnes. This figure is expected to increase to 5.5–6 million tonnes by 2050 (Weckend et al., 2016). Recyclability helps in regaining the materials used in solar panel modules. However, this is a complex process, since the materials used in making solar panel cells need to be extracted after condemnation. Approximately 90% of the solar panel modules use crystalline silicon (c-Si) cells (Metz et al., 2017). The process of regaining materials from c-Si cells at low cost is currently being researched.
- Energy payback is the required time period within which the module is able to produce the same amount of energy that it consumes over its entire life cycle (Desideri & Asdrubali, 2018).
- The quality of service rendered by the manufacturer instils the confidence in users to install solar panels of the said manufacturer. Moreover, in the long run, this helps in building a cordial relationship between the manufacturer and the user, thereby making the product popular and increasing its market share. Better service quality also demands the availability of spare parts as and when required.
- The area covered by the solar panels should be as little as possible in order to generate more amount of power. This will also help in regaining more space for other uses. The material used to make solar panel modules must be environmentally friendly. Silicon is used in approximately 90% of the solar panel modules being used worldwide and is available in abundance in nature.
- The creditability of the supplier is determined by its financial strength, as well as the market reputation. The market reputation further depends upon the reliability of the product and services being offered, along with ISO certification.

9.3 THE HIERARCHICAL STRUCTURE

The criteria discussed can be structured for analysis purposes. Module Efficiency, Power Sorting, Lifetime Energy Yield, PTC Power Rating, and Warranty can be grouped under Performance Criteria. Government Support, Price, Cost per watt can be grouped under Financial Properties. Dimension, Weight, Durability, Corrosion Resistance, Fill Factor, Recyclability, and Energy Payback can be grouped under Technical Characteristics. Service Quality and Spare Parts Availability can be grouped under Customer Support. Area Covered and Material can be grouped under Environment Friendliness. Financial Strength, ISO Certification, and Reliability can be grouped under Supplier Reputation. Figure 9.1 shows the hierarchical structure for solar panel vendor selection.

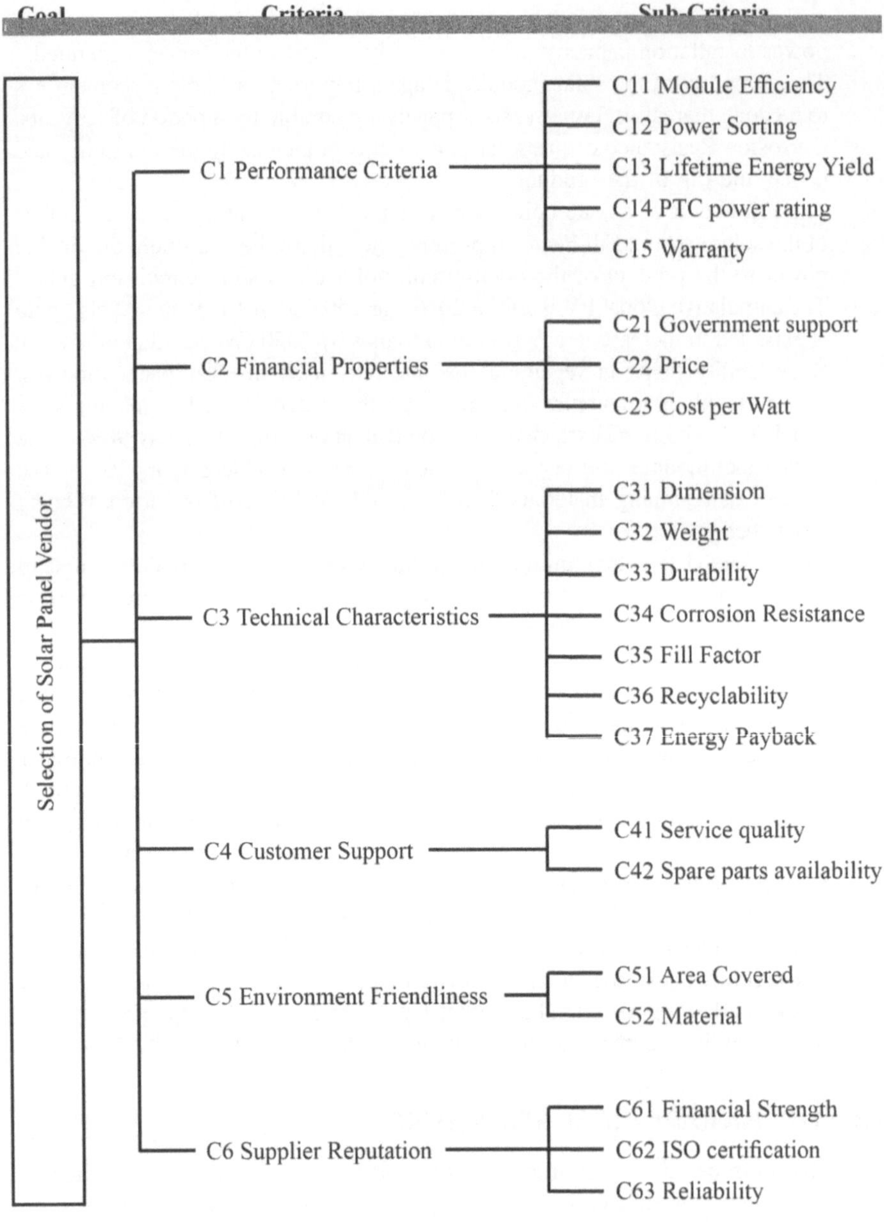

FIGURE 9.1 The Hierarchical Structure of Solar Panel Vendor Selection.

9.4 DETERMINATION OF CRITERIA AND SUB-CRITERIA WEIGHTS

The importance of each of the sub-criteria is different, as they have diverse significance and meaning within the purview of solar panels. As such, the importance of the sub-criteria is determined in terms of weights by employing a suitable method.

Various methods exist for determining weights of the comparative judgment matrix in AHP, such as the eigenvector method, the least-square method, the entropy method, the arithmetic mean method, the geometric mean method, the logistic least squares method (LLSM), and the gradient eigen-root method (GEM) [(Hwang & Yoon, 1981; Sun et al., 2019; Takeda et al., 2020; Xu & Wu 2006)]. The comparative scale used by evaluators is from one to nine in AHP, as per its originator (Saaty, 1980). It has been observed that comparative judgement based on a scale is quite difficult and confusing to comprehend. In order to handle this problem, Saaty's AHP comparative scale was extended by Buckley (Buckley, 1985) by utilizing fuzzy ratios instead of exact ratios. The comparison between two criteria is kept linguistic in nature by asking comparison questions like 'Criteria A has similar importance compared to criteria B'. The fuzzy scale used for the pairwise comparison is shown in Table 9.1.

The decision-making group consisted of manufacturers, experts, and users of solar panels. The data was collected from 10 manufacturers, experts, and users of solar panels in terms of linguistic variables. These decision-makers were selected through random sampling. The weights of the criteria were obtained using the fuzzy AHP method, and subsequently, the average weight was calculated using a geometric mean method, as suggested by Buckley.

Step-1: The data collected from manufacturers are produced in Table 9.2.

TABLE 9.1
Fuzzy Scale for Pairwise Comparison

Intensity of Fuzzy Numbers	Definition of Linguistic Variable	Linguistic Variable	Fuzzy Numbers
9	Absolute importance of criteria A over B	AI	(7,9,9)
7	Demonstrated importance of criteria A over B	DI	(5,7,9)
5	Strong importance of criteria A over B	SI	(3,5,7)
3	Moderate importance of criteria A over B	MI	(1,3,5)
1	Similar importance of criteria A and B	I	(1,1,3)
1/3	Moderate importance of criteria B over A	MI^{-1}	(1/5,1/3,1)
1/5	Strong importance of criteria B over A	SI^{-1}	(1/7,1/5,1/3)
1/7	Demonstrated importance of criteria B over A	DI^{-1}	(1/9,1/7,1/5)
1/9	Absolute importance of criteria B over A	AI^{-1}	(1/9,1/9,1/7)

TABLE 9.2
Data from Manufacturers

	C1	C2	C3	C4	C5	C6
C1	1	I	I	MI^{-1}	I	MI^{-1}
C2		1	MI	I	MI	I
C3			1	MI	MI	MI^{-1}
C4				1	MI	I
C5					1	MI^{-1}
C6						1

Manufacturer 1

	C1	C2	C3	C4	C5	C6
C1	1	MI	I	I	I	MI^{-1}
C2		1	MI	MI	MI	I
C3			1	MI	MI	MI^{-1}
C4				1	MI	I
C5					1	MI^{-1}
C6						1

Manufacturer 2

	C1	C2	C3	C4	C5	C6
C1	1	I	SI	MI	MI	MI^{-1}
C2		1	MI	MI	MI	MI
C3			1	I	MI	I
C4				1	MI	I
C5					1	MI^{-1}
C6						1

Manufacturer 3

	C1	C2	C3	C4	C5	C6
C1	1	MI^{-1}	I	I	I	MI^{-1}
C2		1	MI	MI	I	I
C3			1	I	MI	I
C4				1	MI	MI^{-1}
C5					1	SI^{-1}
C6						1

Manufacturer 4

	C1	C2	C3	C4	C5	C6
C1	1	MI^{-1}	MI	MI^{-1}	MI	I
C2		1	MI	MI	MI	I
C3			1	I	MI	MI^{-1}
C4				1	MI	I
C5					1	MI^{-1}
C6						1

Manufacturer 5

	C1	C2	C3	C4	C5	C6
C1	1	I	I	I	I	MI^{-1}
C2		1	MI	I	I	I
C3			1	I	MI	I
C4				1	MI	MI^{-1}
C5					1	I
C6						1

Manufacturer 6

	C1	C2	C3	C4	C5	C6
C1	1	MI^{-1}	MI^{-1}	I	I	SI^{-1}
C2		1	MI	MI	I	I
C3			1	I	MI	MI^{-1}
C4				1	I	I
C5					1	MI^{-1}
C6						1

Manufacturer 7

	C1	C2	C3	C4	C5	C6
C1	1	MI^{-1}	MI^{-1}	MI^{-1}	I	SI^{-1}
C2		1	MI	MI	MI	I
C3			1	I	MI	I
C4				1	MI	MI^{-1}
C5					1	MI^{-1}
C6						1

Manufacturer 8

	C1	C2	C3	C4	C5	C6
C1	1	MI^{-1}	MI^{-1}	MI^{-1}	I	I
C2		1	MI	MI	MI	MI
C3			1	I	MI	MI^{-1}
C4				1	I	I
C5					1	I
C6						1

Manufacturer 9

	C1	C2	C3	C4	C5	C6
C1	1	I	I	MI^{-1}	I	SI^{-1}
C2		1	I	I	MI	I
C3			1	I	MI	I
C4				1	MI	I
C5					1	MI^{-1}
C6						1

Manufacturer 10

Step-2: The next step is to convert these data into the intensity of fuzzy numbers, as already defined in Table 9.1. The converted data is shown in Table 9.3.

Step-3: After capturing the intensity of fuzzy numbers based on data provided by manufacturers, the synthetic pairwise comparison matrix needs to be obtained using the geometric mean method, as suggested by Buckley:

$$\tilde{a}_{ij} = \left(\tilde{a}_{ij}^{1} \otimes \tilde{a}_{ij}^{2} \otimes \tilde{a}_{ij}^{3} \otimes \tilde{a}_{ij}^{4} \otimes \tilde{a}_{ij}^{5} \otimes \tilde{a}_{ij}^{6} \otimes \tilde{a}_{ij}^{7} \otimes \tilde{a}_{ij}^{8} \otimes \tilde{a}_{ij}^{9} \otimes \tilde{a}_{ij}^{10}\right)^{\frac{1}{10}} \quad (9.1)$$

Let us study how to find the value of a_{12}. Each of the elements a_{12} of the manufacturer 1–10 must be converted with their fuzzy numbers, as per Table 9.1, and applied in the equation above:

$$\tilde{a}_{12} = \left((1,1,3) \otimes (1,3,5) \otimes (1,1,3) \otimes \left(\tfrac{1}{5},\tfrac{1}{3},1\right) \otimes \left(\tfrac{1}{5},\tfrac{1}{3},1\right) \otimes (1,1,3) \otimes \left(\tfrac{1}{5},\tfrac{1}{3},1\right) \otimes \left(\tfrac{1}{5},\tfrac{1}{3},1\right) \otimes \left(\tfrac{1}{5},\tfrac{1}{3},1\right) \otimes (1,1,3)\right)^{\frac{1}{10}}$$

$$\tilde{a}_{12} = \left(\left(1 \times 1 \times 1 \times \tfrac{1}{5} \times \tfrac{1}{5} \times 1 \times \tfrac{1}{5} \times \tfrac{1}{5} \times \tfrac{1}{5} \times 1\right)^{\frac{1}{10}}, \left(1 \times 3 \times 1 \times \tfrac{1}{3} \times \tfrac{1}{3} \times 1 \times \tfrac{1}{3} \times \tfrac{1}{3} \times \tfrac{1}{3} \times 1\right)^{\frac{1}{10}}, (3 \times 5 \times 3 \times 1 \times 1 \times 3 \times 1 \times 1 \times 1 \times 3)^{\frac{1}{10}}\right)$$

$$\tilde{a}_{12} = (0.447,\ 0.644,\ 1.823)$$

In a similar way, other fuzzy elements may be calculated. The pairwise comparison matrix of the ten representatives of the manufacturer will be constructed, as per Table 9.4.

Step-4: To obtain the fuzzy weights of Criteria 1 for the manufacturer group, the following equations may be used:

$$\begin{aligned}\tilde{f}_1 &= \left(\tilde{a}_{11} \otimes \tilde{a}_{12} \otimes \tilde{a}_{13} \otimes \tilde{a}_{14} \otimes \tilde{a}_{15} \otimes \tilde{a}_{16}\right)^{\frac{1}{6}} \\ &= \left((1 \times 0.447 \times 0.689 \times 0.447 \times 1 \times 0.258)^{\frac{1}{6}}, (1 \times 0.644 \times \ldots \times 0.375)^{\frac{1}{6}}, (1 \times 1.823 \times \ldots \times 1)^{\frac{1}{6}}\right) \\ &= (0.573,\ 0.753,\ 1.735)\end{aligned} \quad (9.2)$$

TABLE 9.3
Conversion of Data from Manufacturers to Intensity of Fuzzy Numbers

	C1	C2	C3	C4	C5	C6
C1	1	1	1	1/3	1	1/3
C2	1	1	3	1	3	1
C3	1	1/3	1	3	3	1/3
C4	3	1	1/3	1	3	1
C5	1	1/3	1/3	1/3	1	1/5
C6	3	1	3	1	5	1

Manufacturer 1

	C1	C2	C3	C4	C5	C6
C1	1	3	1	1	1	1/3
C2	1/3	1	3	3	3	1
C3	1	1/3	1	3	3	1/3
C4	1	1/3	1/3	1	3	1
C5	1	1/3	1/3	1/3	1	1/3
C6	3	1	3	1	3	1

Manufacturer 2

	C1	C2	C3	C4	C5	C6
C1	1	1	5	3	3	1/3
C2	1	1	3	3	3	3
C3	1/5	1/3	1	1	3	1
C4	1/3	1/3	1	1	3	1
C5	1/3	1/3	1/3	1/3	1	1/3
C6	3	1/3	1	1	3	1

Manufacturer 3

	C1	C2	C3	C4	C5	C6
C1	1	1/3	1	1	1	1/3
C2	3	1	3	3	1	1
C3	1	1/3	1	1	3	1
C4	1	1/3	1	1	3	1/3
C5	1	1	1/3	1/3	1	1/5
C6	3	1	1	3	5	1

Manufacturer 4

	C1	C2	C3	C4	C5	C6
C1	1	1/3	3	1/3	3	1
C2	3	1	3	3	3	1
C3	1/3	1/3	1	1	3	1/3
C4	3	1/3	1	1	3	1
C5	1/3	1/3	1/3	1/3	1	1/3
C6	1	1	3	1	3	1

Manufacturer 5

	C1	C2	C3	C4	C5	C6
C1	1	1	1	1	1	1/3
C2	1	1	3	1	1	1
C3	1	1/3	1	1	3	1
C4	1	1	1	1	3	1/3
C5	1	1	1/3	1/3	1	1
C6	3	1	1	3	1	1

Manufacturer 6

	C1	C2	C3	C4	C5	C6
C1	1	1/3	1/3	1	1	1/5
C2	3	1	3	3	1	1
C3	3	1/3	1	1	3	1/3
C4	1	1/3	1	1	1	1
C5	1	1	1/3	1	1	1/3
C6	5	1	3	1	3	1

Manufacturer 7

	C1	C2	C3	C4	C5	C6
C1	1	1/3	1/3	1/3	1	1/5
C2	3	1	3	3	3	1
C3	3	1/3	1	1	3	1
C4	3	1/3	1	1	3	1/3
C5	1	1/3	1/3	1/3	1	1/3
C6	5	1	1	3	3	1

Manufacturer 8

	C1	C2	C3	C4	C5	C6
C1	1	1/3	1/3	1/3	1	1
C2	3	1	3	3	3	3
C3	3	1/3	1	1	3	1/3
C4	3	1/3	1	1	1	1
C5	1	1/3	1/3	1	1	1
C6	1	1/3	3	1	1	1

Manufacturer 9

	C1	C2	C3	C4	C5	C6
C1	1	1	1	1/3	1	1/3
C2	1	1	1	1	3	1
C3	1	1	1	1	3	1
C4	3	1	1	1	3	1
C5	1	1/3	1/3	1/3	1	1/3
C6	1	1	1	1/3	1	1/3

Manufacturer 10

TABLE 9.4
Pairwise Comparison Matrix of Manufacturer Representatives

	C1	C2	C3	C4	C5	C6
C1	1	(0.447,0.644,1.823)	(0.689,0.943,2.472)	(0.447,0.644,1.823)	(1.000,1.246,3.323)	(0.258,0.375,1.000)
C2	(0.851,1.552,3.470)	1	(1.000,2.688,4.751)	(0.851,1.732,3.652)	(1.000,2.158,4.290)	(0.725,1.000,2.667)
C3	(0.701,1.061,2.515)	(0.235,0.372,1.116)	1	(1.000,1.246,3.323)	(1.000,3.000,5.000)	(0.447,0.577,1.732)
C4	(0.851,1.552,3.470)	(0.381,0.577,1.633)	(0.725,0.803,2.408)	1	(1.000,2.408,4.514)	(0.617,0.719,2.158)
C5	(0.725,0.803,2.408)	(0.324,0.463,1.390)	(0.200,0.333,1.000)	(0.276,0.415,1.246)	1	(0.258,0.375,1.000)
C6	(1.246,2.667,4.829)	(0.725,1.000,2.667)	(1.000,1.732,3.873)	(1.000,1.390,3.497)	(1.246,2.667,4.829)	1

Similarly, the fuzzy weights of other criteria for the manufacturers group may be calculated as follows:

$\tilde{f}_2 = (0.898, 1.581, 2.972); \tilde{f}_3 = (0.647, 0.974, 2.079); \tilde{f}_4 = (0.725, 1.037, 2.259);$
$\tilde{f}_5 = (0.387, 0.518, 1.269); \tilde{f}_6 = (1.020, 1.606, 3.073)$

Step-5: At this point, the succeeding equation is used to calculate the weight of first criterion:

$$\begin{aligned}\tilde{w}_1 &= \frac{\tilde{f}_1}{\left(\tilde{f}_1 \oplus \tilde{f}_2 \oplus \tilde{f}_3 \oplus \tilde{f}_4 \oplus \tilde{f}_5 \oplus \tilde{f}_6\right)} \\ &= \frac{(0.573, 0.753, 1.735)}{(0.573, 0.753, 1.735) \oplus (0.898, 1.581, 2.972) \oplus (0.647, 0.974, 2.079) \oplus (0.725, 1.037, 2.259) \oplus (0.387, 0.518, 1.269) \oplus (1.020, 1.606, 3.073)} \\ &= \left(\frac{0.573}{4.250}, \frac{0.753}{6.468}, \frac{1.735}{13.387}\right) = (0.135, 0.116, 0.130)\end{aligned} \tag{9.3}$$

Similarly, the weight of all of the criteria may be calculated:

$$\tilde{w}_2 = (0.211, 0.244, 0.222);\ \tilde{w}_3 = (0.152, 0.151, 0.155);\ \tilde{w}_4 = (0.171, 0.160, 0.169);$$
$$\tilde{w}_5 = (0.091, 0.080, 0.095);\ \tilde{w}_6 = (0.240, 0.248, 0.230)$$

Step-6: To calculate the Best Non-fuzzy Performance (BNP) of first criteria using the Centre of Area (CoA) method, the following equations may be accomplished:

$$BNP_{w1} = \frac{(U_{w1} - L_{w1}) + (M_{w1} - L_{w1})}{3} + L_{w1} \tag{9.4}$$

$$BNP_{w1} = \frac{(0.135 - 0.116) + (0.130 - 0.116)}{3} + 0.116 = 0.127$$

Similarly, the BNP of all criteria may be calculated as follows:

$$BNP_{w2} = 0.226;\ BNP_{w3} = 0.153;\ BNP_{w4} = 0.167;\ BNP_{w5} = 0.089;\ BNP_{w6} = 0.239$$

Steps 1–6 will be repeated for determining BNP values of sub-criteria for all six criteria. In this case, the computation is achieved for the sub-criteria of the first criterion, Performance Criteria.

Step-1: The data collected from manufacturers for performance criteria are produced in Table 9.5.

TABLE 9.5
Data from Manufacturers for Performance Criteria

	C11	C12	C13	C14	C15
C11	1	MI	MI^{-1}	I	I
C12		1	MI^{-1}	I	I
C13			1	MI	I
C14				1	MI
C15					1

Manufacturer 1

	C11	C12	C13	C14	C15
C11	1	MI	I	I	MI
C12		1	I	I	MI
C13			1	MI	MI
C14				1	MI
C15					1

Manufacturer 2

	C11	C12	C13	C14	C15
C11	1	I	I	MI	MI
C12		1	MI	I	MI
C13			1	I	I
C14				1	MI
C15					1

Manufacturer 3

	C11	C12	C13	C14	C15
C11	1	MI	I	I	SI
C12		1	MI	I	I
C13			1	I	I
C14				1	MI
C15					1

Manufacturer 4

	C11	C12	C13	C14	C15
C11	1	I	I	I	MI
C12		1	I	MI	MI
C13			1	I	I
C14				1	I
C15					1

Manufacturer 5

	C11	C12	C13	C14	C15
C11	1	I	I	I	MI
C12		1	MI	I	I
C13			1	I	MI
C14				1	MI
C15					1

Manufacturer 6

	C11	C12	C13	C14	C15
C11	1	I	I	I	I
C12		1	MI	MI	I
C13			1	I	MI
C14				1	I
C15					1

Manufacturer 7

	C11	C12	C13	C14	C15
C11	1	I	I	MI	MI
C12		1	I	MI	MI
C13			1	I	I
C14				1	I
C15					1

Manufacturer 8

	C11	C12	C13	C14	C15
C11	1	MI^{-1}	MI^{-1}	MI^{-1}	I
C12		1	MI	MI	MI
C13			1	I	MI
C14				1	I
C15					1

Manufacturer 9

	C11	C12	C13	C14	C15
C11	1	I	I	MI^{-1}	I
C12		1	I	I	MI
C13			1	I	MI
C14				1	MI
C15					1

Manufacturer 10

Step-2: The data collected from manufacturers for performance criteria in Table 9.5 is converted to intensity of fuzzy numbers, as shown in Table 9.6.

Step-3: Based on the intensity of fuzzy numbers provided by manufacturers for performance criteria, the synthetic pairwise comparison matrix is obtained using the geometric mean method, as shown in Table 9.7.

TABLE 9.6
Conversion of Data for Performance Criteria from Manufacturers to Intensity of Fuzzy Numbers

	C11	C12	C13	C14	C15
C11	1	3	1/3	1	1
C12		1	1/3	1	1
C13			1	3	1
C14				1	3
C15					1

Manufacturer 1

	C11	C12	C13	C14	C15
C11	1	3	1	1	3
C12		1	1	1	3
C13			1	3	3
C14				1	3
C15					1

Manufacturer 2

	C11	C12	C13	C14	C15
C11	1	1	1	3	3
C12		1	3	1	3
C13			1	1	1
C14				1	3
C15					1

Manufacturer 3

	C11	C12	C13	C14	C15
C11	1	3	1	1	5
C12		1	3	1	1
C13			1	1	1
C14				1	3
C15					1

Manufacturer 4

	C11	C12	C13	C14	C15
C11	1	1	1	1	3
C12		1	1	3	3
C13			1	1	1
C14				1	1
C15					1

Manufacturer 5

	C11	C12	C13	C14	C15
C11	1	1	1	1	3
C12		1	3	1	1
C13			1	1	3
C14				1	3
C15					1

Manufacturer 6

	C11	C12	C13	C14	C15
C11	1	1	1	1	1
C12		1	3	3	1
C13			1	1	3
C14				1	1
C15					1

Manufacturer 7

	C11	C12	C13	C14	C15
C11	1	1	1	3	3
C12		1	1	3	3
C13			1	1	1
C14				1	1
C15					1

Manufacturer 8

	C11	C12	C13	C14	C15
C11	1	3^{-1}	3^{-1}	3^{-1}	1
C12		1	3	3	3
C13			1	1	3
C14				1	1
C15					1

Manufacturer 9

	C11	C12	C13	C14	C15
C11	1	1	1	3^{-1}	1
C12		1	1	1	3
C13			1	1	3
C14				1	3
C15					1

Manufacturer 10

TABLE 9.7
Pairwise Comparison Matrix of Performance Criteria Using Data Provided by Manufacturers

	C11	C12	C13	C14	C15
C11	1	(0.851,1.246,3.133)	(0.725,0.803,2.408)	(0.725,1.000,2.667)	(1.116,2.034,4.215)
C12	(0.617,0.803,2.271)	1	(0.851,1.552,3.470)	(1.000,1.552,3.680)	(1.000,1.933,4.076)
C13	(1.000,1.246,3.323)	(0.447,0.644,1.823)	1	(1.000,1.246,3.323)	(1.000,1.732,3.873)
C14	(0.725,1.000,2.667)	(0.525,0.644,1.933)	(0.725,0.803,2.408)	1	(1.000,1.933,4.076)
C15	(0.368,0.492,1.390)	(0.381,0.517,1.552)	(0.447,0.577,1.732)	(0.381,0.517,1.552)	1

Step-4: The fuzzy weights of the sub-criteria 'module efficiency' of 'performance criteria' for manufacturer groups are depicted below:

$$\tilde{f}_{11} = \left((1 \times 0.851 \times 0.725 \times 0.725 \times 1.116)^{\frac{1}{5}}, (1 \times 1.246 \times \ldots \times 2.034)^{\frac{1}{5}},\right.$$
$$\left.(1 \times 3.133 \times \ldots \times 4.215)^{\frac{1}{5}}\right)$$
$$= (0.870, 1.153, 2.431)$$

Similarly, $\tilde{f}_{12}$ = (0.879, 1.302, 2.597); $\tilde{f}_{13}$ = (0.851, 1.116, 2.390); $\tilde{f}_{14}$ = (0.773, 1.000, 2.192); $\tilde{f}_{15}$ = (0.474, 0.597, 1.421)

Step-5: The calculation of the weight of sub-criteria of the first criterion is shown below:

$$\tilde{w}_{11} = \frac{(00.870, 1.153, 2.431)}{\begin{array}{l}(0.870, 1.153, 2.431) \oplus (0.879, 1.302, 2.597) \\ \oplus (0.851, 1.116, 2.390) \oplus (0.773, 1.000, 2.192) \\ \oplus (0.474, 0.597, 1.421)\end{array}}$$

$$\tilde{w}_{11} = (0.226, 0.223, 0.220)$$

Similarly,

$$\tilde{w}_{12} = (0.229, 0.252, 0.235); \; \tilde{w}_{13} = (0.221, 0.216, 0.217);$$
$$\tilde{w}_{14} = (0.201, 0.194, 0.199);$$
$$\tilde{w}_{15} = (0.123, 0.116, 0.129)$$

Step-6: The Best Non-fuzzy Performance (BNP) of the sub-criteria of the first criteria using the Centre of Area (CoA) method can be calculated as shown:

$$BNP_{w11} = \frac{(U_{w11} - L_{w11}) + (M_{w11} - L_{w11})}{3} + L_{w11}$$

$$BNP_{w11} = 0.223; \; BNP_{w12} = 0.239; \; BNP_{w13} = 0.218; \; BNP_{w14} = 0.198;$$
$$BNP_{w15} = 0.123$$

Steps 1–6 need to be repeated for expert and user groups, as well as for all criteria and sub-criteria, for calculating BNP values.

9.5 CRITICAL WEIGHT OF MANUFACTURERS, EXPERTS, AND USERS

The weight of criteria and sub-criteria for the manufacturers, experts, and users group are listed in Tables 9.8, 9.9, and 9.10 respectively.

9.6 EVALUATOR'S ESTIMATION OF THE PERFORMANCE OF THE ALTERNATIVES

Five evaluators were given the task of ranking the five available alternatives for solar modules. The evaluators were required to estimate their subjective judgement

TABLE 9.8
Weight of Criteria and Sub-Criteria for Manufacturers Group

	Criteria and Sub-Criteria	Local Weights	Overall Weights	BNP
C1	**Performance Criteria**	(0.135,0.116,0.13)		0.127
C11	Module Efficiency	(0.226,0.223,0.22)	(0.031,0.026,0.029)	0.028
C12	Power Sorting	(0.229,0.252,0.235)	(0.031,0.029,0.031)	0.030
C13	Lifetime Energy Yield	(0.221,0.216,0.217)	(0.03,0.025,0.028)	0.028
C14	PTC Power Rating	(0.201,0.194,0.199)	(0.027,0.023,0.026)	0.025
C15	Warranty	(0.123,0.116,0.129)	(0.017,0.013,0.017)	0.016
C2	**Financial Properties**	(0.211,0.244,0.222)		0.226
C21	Government Support	(0.351,0.333,0.34)	(0.074,0.081,0.075)	0.077
C22	Price	(0.333,0.358,0.345)	(0.07,0.087,0.077)	0.078
C23	Cost per Watt	(0.316,0.309,0.316)	(0.067,0.076,0.07)	0.071
C3	**Technical Characteristics**	(0.152,0.151,0.155)		0.153
C31	Dimension	(0.072,0.064,0.075)	(0.011,0.01,0.012)	0.011
C32	Weight	(0.092,0.077,0.093)	(0.014,0.012,0.014)	0.013
C33	Durability	(0.127,0.142,0.139)	(0.019,0.021,0.022)	0.021
C34	Corrosion Resistance	(0.111,0.096,0.11)	(0.017,0.014,0.017)	0.016
C35	Fill Factor	(0.196,0.166,0.176)	(0.03,0.025,0.027)	0.027
C36	Recyclability	(0.193,0.244,0.21)	(0.029,0.037,0.033)	0.033
C37	Energy Payback	(0.209,0.21,0.197)	(0.032,0.032,0.031)	0.031
C4	**Customer Support**	(0.171,0.16,0.169)		0.167
C41	Service Quality	(0.635,0.67,0.635)	(0.108,0.108,0.107)	0.108
C42	Spare Parts Availability	(0.365,0.33,0.365)	(0.062,0.053,0.062)	0.059
C5	**Environment Friendliness**	(0.091,0.08,0.095)		0.089
C51	Area Covered	(0.384,0.354,0.384)	(0.035,0.028,0.036)	0.033
C52	Material	(0.616,0.646,0.616)	(0.056,0.052,0.058)	0.055
C6	**Supplier Reputation**	(0.24,0.248,0.23)		0.239
C61	Financial Strength	(0.511,0.604,0.524)	(0.123,0.15,0.12)	0.131
C62	ISO Certification	(0.221,0.175,0.208)	(0.053,0.043,0.048)	0.048
C63	Reliability	(0.268,0.221,0.268)	(0.064,0.055,0.061)	0.060

TABLE 9.9
Weights of Criteria and Sub-Criteria for the Experts Group

	Criteria and Sub-Criteria	Local Weights	Overall Weights	BNP
C1	**Performance Criteria**	(0.191,0.206,0.199)		0.199
C11	Module Efficiency	(0.262,0.304,0.27)	(0.05,0.063,0.054)	0.055
C12	Power Sorting	(0.19,0.196,0.195)	(0.036,0.04,0.039)	0.039
C13	Lifetime Energy Yield	(0.196,0.184,0.192)	(0.037,0.038,0.038)	0.038
C14	PTC Power Rating	(0.216,0.192,0.203)	(0.041,0.04,0.04)	0.040
C15	Warranty	(0.137,0.124,0.14)	(0.026,0.026,0.028)	0.027
C2	**Financial Properties**	(0.074,0.069,0.079)		0.074
C21	Government Support	(0.454,0.518,0.463)	(0.034,0.036,0.036)	0.035
C22	Price	(0.266,0.258,0.275)	(0.02,0.018,0.022)	0.020
C23	Cost per Watt	(0.28,0.223,0.262)	(0.021,0.015,0.021)	0.019
C3	**Technical Characteristics**	(0.176,0.161,0.17)		0.169
C31	Dimension	(0.104,0.083,0.1)	(0.018,0.013,0.017)	0.016
C32	Weight	(0.097,0.076,0.095)	(0.017,0.012,0.016)	0.015
C33	Durability	(0.096,0.097,0.103)	(0.017,0.016,0.018)	0.017
C34	Corrosion Resistance	(0.139,0.103,0.123)	(0.024,0.017,0.021)	0.021
C35	Fill Factor	(0.16,0.151,0.155)	(0.028,0.024,0.026)	0.026
C36	Recyclability	(0.218,0.264,0.226)	(0.038,0.042,0.038)	0.040
C37	Energy Payback	(0.185,0.226,0.198)	(0.033,0.036,0.034)	0.034
C4	**Customer Support**	(0.229,0.234,0.22)		0.228
C41	Service Quality	(0.704,0.749,0.704)	(0.161,0.175,0.155)	0.164
C42	Spare Parts Availability	(0.296,0.251,0.296)	(0.068,0.059,0.065)	0.064
C5	**Environment Friendliness**	(0.122,0.116,0.126)		0.121
C51	Area Covered	(0.5,0.5,0.5)	(0.061,0.058,0.063)	0.061
C52	Material	(0.5,0.5,0.5)	(0.061,0.058,0.063)	0.061
C6	**Supplier Reputation**	(0.209,0.214,0.207)		0.210
C61	Financial Strength	(0.427,0.504,0.446)	(0.089,0.108,0.092)	0.096
C62	ISO Certification	(0.263,0.225,0.257)	(0.055,0.048,0.053)	0.052
C63	Reliability	(0.309,0.271,0.297)	(0.065,0.058,0.061)	0.061

for each alternative against each of the sub-criteria. Since individual evaluators may have different views on judging an alternative of solar modules, with respect to a sub-criterion, a range must be defined within which a particular linguistic variable may lie. Table 9.11 enlists the scope for linguistic variables of Very Low (VL), Low (L), Medium (M), High (H), and Very High (VH), as provided by the evaluators. Table 9.12 provides subjective judgement given by the evaluators for each of the alternatives against each of the sub-criterion.

To understand the calculation of the fuzzy performance value of each alternative of solar modules as provided by evaluators, let us consider the case of Solar Module Vendor 1 against C11-Module Efficiency. From Table 9.12, the linguistic variables

TABLE 9.10
Weights of Criteria and Sub-Criteria for the Users Group

	Criteria and Sub-Criteria	Local Weights	Overall Weights	BNP
C1	**Performance Criteria**	(0.289,0.335,0.287)		0.304
C11	Module Efficiency	(0.312,0.376,0.322)	(0.09,0.126,0.092)	0.103
C12	Power Sorting	(0.141,0.126,0.142)	(0.041,0.042,0.041)	0.041
C13	Lifetime Energy Yield	(0.183,0.173,0.184)	(0.053,0.058,0.053)	0.055
C14	PTC Power Rating	(0.172,0.148,0.167)	(0.05,0.05,0.048)	0.049
C15	Warranty	(0.193,0.177,0.184)	(0.056,0.059,0.053)	0.056
C2	**Financial Properties**	(0.072,0.059,0.07)		0.067
C21	Government Support	(0.504,0.562,0.507)	(0.036,0.033,0.036)	0.035
C22	Price	(0.184,0.157,0.188)	(0.013,0.009,0.013)	0.012
C23	Cost per Watt	(0.313,0.281,0.304)	(0.022,0.017,0.021)	0.020
C3	**Technical Characteristics**	(0.161,0.156,0.165)		0.161
C31	Dimension	(0.11,0.091,0.106)	(0.018,0.014,0.018)	0.016
C32	Weight	(0.161,0.128,0.146)	(0.026,0.02,0.024)	0.023
C33	Durability	(0.162,0.159,0.161)	(0.026,0.025,0.027)	0.026
C34	Corrosion Resistance	(0.082,0.066,0.084)	(0.013,0.01,0.014)	0.013
C35	Fill Factor	(0.113,0.086,0.106)	(0.018,0.013,0.017)	0.016
C36	Recyclability	(0.141,0.151,0.146)	(0.023,0.024,0.024)	0.023
C37	Energy Payback	(0.231,0.318,0.25)	(0.037,0.05,0.041)	0.043
C4	**Customer Support**	(0.233,0.231,0.226)		0.230
C41	Service Quality	(0.721,0.769,0.721)	(0.168,0.177,0.163)	0.169
C42	Spare Parts Availability	(0.279,0.231,0.279)	(0.065,0.053,0.063)	0.061
C5	**Environment Friendliness**	(0.096,0.081,0.096)		0.091
C51	Area Covered	(0.689,0.739,0.689)	(0.066,0.06,0.066)	0.064
C52	Material	(0.311,0.261,0.311)	(0.03,0.021,0.03)	0.027
C6	**Supplier Reputation**	(0.149,0.138,0.155)		0.147
C61	Financial Strength	(0.36,0.362,0.357)	(0.054,0.05,0.056)	0.053
C62	ISO Certification	(0.261,0.233,0.259)	(0.039,0.032,0.04)	0.037
C63	Reliability	(0.38,0.404,0.384)	(0.057,0.056,0.06)	0.057

TABLE 9.11
Evaluator's Range for Linguistic Variables

Evaluator	Very Low (VL)	Low (L)	Medium (M)	High (H)	Very High (VH)
1	(0,0,40)	(40,50,60)	(60,70,80)	(80,85,90)	(90,100,100)
2	(0,0,20)	(20,30,40)	(35,45,70)	(70,80,90)	(85,100,100)
3	(0,0,25)	(10,30,50)	(30,50,70)	(65,75,85)	(80,100,100)
4	(0,0,30)	(15,35,55)	(40,55,70)	(60,75,90)	(85,100,100)
5	(0,0,15)	(15,30,45)	(45,60,75)	(75,80,90)	(90,100,100)

TABLE 9.12
Linguistic Values by Evaluators

		Solar Module Vendor 1					Solar Module Vendor 2					Solar Module Vendor 3					Solar Module Vendor 4					Solar Module Vendor 5				
		E1	E2	E3	E4	E5	E1	E2	E3	E4	E5	E1	E2	E3	E4	E5	E1	E2	E3	E4	E5	E1	E2	E3	E4	E5
C11	Module Efficiency	M	M	M	L	M	H	M	M	H	M	L	M	M	L	M	M	L	M	L	M	M	L	M	L	M
C12	Power Sorting	H	L	M	M	H	M	M	H	L	M	M	M	L	M	H	L	M	L	L	M	L	M	L	M	L
C13	Lifetime Energy Yield	L	L	M	L	M	M	L	M	M	L	M	L	M	L	M	L	L	M	L	M	L	L	M	L	L
C14	PTC Power Rating	M	M	L	M	L	M	M	L	M	L	H	L	M	H	L	L	L	L	VL	VL	L	VL	L	L	L
C15	Warranty	M	M	M	L	L	M	M	L	L	M	M	L	M	L	M	L	L	M	L	L	L	L	L	L	L
C21	Government Support	H	H	M	H	H	M	H	H	H	H	H	M	M	H	M	H	M	H	M	H	M	M	L	M	L
C22	Price	M	M	L	M	L	L	L	M	VL	L	M	M	L	M	L	L	M	L	M	L	M	L	L	M	L
C23	Cost per Watt	M	L	M	M	M	M	M	L	M	L	M	L	M	L	M	M	L	M	L	M	M	M	L	M	L
C31	Dimension	H	H	M	M	M	H	H	H	M	H	M	L	M	M	H	H	H	H	M	H	H	M	M	M	H
C32	Weight	H	M	M	M	H	H	H	M	H	M	M	M	L	M	M	L	M	L	M	L	L	L	M	M	L
C33	Durability	M	M	M	L	M	M	M	H	M	M	M	H	M	H	M	M	M	L	L	M	L	L	M	L	M
C34	Corrosion Resistance	M	M	L	M	L	M	M	H	H	M	M	L	L	M	L	M	L	L	M	L	L	L	L	M	L
C35	Fill Factor	M	H	M	H	M	H	H	M	H	H	L	M	M	L	M	M	L	M	L	M	M	L	M	L	M
C36	Recyclability	L	M	M	M	L	H	H	M	H	M	L	M	M	H	H	L	M	L	M	L	L	M	L	M	L
C37	Energy Payback	M	M	L	M	L	M	M	M	H	M	M	M	L	M	L	M	M	M	M	L	M	L	L	M	L
C41	Service Quality	M	M	M	L	M	M	M	H	M	M	M	M	H	H	M	M	H	M	L	M	M	L	M	L	M
C42	Spare Parts Availability	H	M	M	H	M	M	H	H	M	H	H	M	M	H	M	L	M	M	L	M	H	M	M	L	M
C51	Area Covered	H	M	M	M	M	M	H	M	H	H	VH	M	H	M	H	M	M	L	M	M	L	M	L	M	L
C52	Material	L	M	M	M	L	L	M	M	L	L	M	M	M	VH	VH	M	M	M	L	L	L	M	M	L	M
C61	Financial Strength	M	H	M	H	H	H	H	M	H	M	M	L	M	M	L	M	L	L	M	L	M	L	M	H	L
C62	ISO certification	M	M	L	M	M	H	M	H	M	H	M	M	L	M	M	M	M	L	L	M	M	M	L	M	M
C63	Reliability	M	M	H	M	H	M	H	H	M	M	M	M	L	M	L	L	M	L	M	VL	M	M	L	M	L

TABLE 9.13
Fuzzy Value of Each Linguistic Variable

E1	E2	E3	E4	E5
M	M	M	L	M
(60,70,80)	(35,45,70)	(30,50,70)	(15,35,55)	(45,60,75)

provided by each evaluator E1, E2, E3, E4, E5 are M, M, M, L, M respectively. The fuzzy value of each linguistic variable is described, as per Table 9.13.

The Average Fuzzy Performance (AFP) value for alternative Solar Module Vendor 1 against C11- Module Efficiency can be calculated using above fuzzy values:

$$AFP_{C11}^{SM1} = \left(\frac{60 + 35 + 30 + 15 + 45}{5}, \frac{70 + 45 + 50 + 35 + 60}{5}, \frac{80 + 70 + 70 + 55 + 75}{5} \right)$$

$$AFP_{C11}^{SM1} = (37, 52, 70)$$

Similarly, all other AFP values can be calculated and are listed in Table 9.14.

9.7 RANKING OF THE ALTERNATIVES

The three decision-making groups, consisting of manufacturers, experts, and users, have provided their inputs in the form of weights for each of the sub-criterion, as shown in Table 9.8, 9.9, and 9.10 respectively. Based on the weights of decision-making groups and the ratings provided by evaluators for each of the sub-criteria, the BNP value can now be calculated for solar module alternatives.

The ranking value to be used for identifying the BNP values for the manufacturers group is calculated as follows:

$$\begin{aligned} R_{m11} &= (LR_{m11}, MR_{m11}, UR_{m11}) = \left(\sum_{j=1}^{22} LE_j \times LM_j, \sum_{j=1}^{22} ME_j \times MM_j, \sum_{j=1}^{22} UE_j \times UM_j \right) \\ &= [(37 \times 0.031 + \ldots + 55 \times 0.064), (52 \times 0.026 + \ldots + 65 \times 0.055), (70 \times 0.029 \\ &\quad + \ldots + 79 \times 0.061)] \\ &= (43.86, 57.35, 72.44) \end{aligned} \tag{9.5}$$

The BNP value of Solar Module Vendor 1 can be calculated using the above-mentioned ranking values for the manufacturers group:

TABLE 9.14
Average Values of Linguistic Variables for Each Alternative against Each Sub-Criterion

	Sub-Criteria	Solar Module Vendor 1	Solar Module Vendor 2	Solar Module Vendor 3	Solar Module Vendor 4	Solar Module Vendor 5
C11	Module Efficiency	(37,52,70)	(50,63,79)	(33,48,66)	(34,49,64)	(34,49,64)
C12	Power Sorting	(49,60,72)	(44,57,73)	(44,56,72)	(29,44,62)	(28,42,59)
C13	Lifetime Energy Yield	(30,45,60)	(33,47,61)	(34,49,64)	(30,45,60)	(24,39,54)
C14	PTC Power Rating	(32,46,63)	(32,46,63)	(41,54,67)	(14,22,39)	(16,29,46)
C15	Warranty	(31,46,64)	(33,48,66)	(34,49,64)	(24,39,54)	(20,35,50)
C21	Government Support	(63,74,86)	(66,76,87)	(50,63,79)	(59,68,81)	(32,46,63)
C22	Price	(32,46,63)	(21,32,49)	(32,46,63)	(28,42,59)	(29,43,57)
C23	Cost per Watt	(39,53,67)	(32,46,63)	(34,49,64)	(34,49,64)	(32,46,63)
C31	Dimension	(53,66,79)	(66,75,85)	(45,57,70)	(66,75,85)	(52,63,78)
C32	Weight	(52,63,78)	(57,70,83)	(38,52,69)	(28,42,59)	(29,43,57)
C33	Durability	(37,52,70)	(49,61,76)	(53,67,81)	(33,48,66)	(30,45,60)
C34	Corrosion Resistance	(32,46,63)	(53,65,80)	(29,43,57)	(29,43,57)	(25,39,53)
C35	Fill Factor	(53,67,81)	(63,74,86)	(33,48,66)	(34,49,64)	(34,49,64)
C36	Recyclability	(32,46,63)	(57,70,83)	(48,60,76)	(28,42,59)	(28,42,59)
C37	Energy Payback	(32,46,63)	(46,60,77)	(32,46,63)	(36,50,67)	(29,43,57)
C41	Service Quality	(37,52,70)	(49,61,76)	(53,65,80)	(44,59,74)	(34,49,64)
C42	Spare Parts Availability	(50,63,79)	(62,72,83)	(50,63,79)	(33,48,66)	(41,55,72)
C51	Area Covered	(46,59,75)	(59,71,84)	(61,71,83)	(38,52,69)	(28,42,59)
C52	Material	(32,46,63)	(27,42,60)	(60,73,84)	(31,46,64)	(33,48,66)
C61	Financial Strength	(59,71,84)	(57,70,83)	(33,47,61)	(29,43,57)	(37,51,65)
C62	ISO Certification	(38,52,69)	(59,68,81)	(38,52,69)	(33,48,66)	(38,52,69)
C63	Reliability	(55,65,79)	(56,68,80)	(32,46,63)	(25,36,53)	(32,46,63)

TABLE 9.15
Final Ranking of Solar Modules Based on Decision Groups

	Manufacturers Group		Experts Group		Users Group		Compromised	
Alternative	BNP	Ranking	BNP	Ranking	BNP	Ranking	BNP	Ranking
Solar Module Vendor 1	57.884	2	56.742	3	55.828	3	56.818	2
Solar Module Vendor 2	61.315	1	62.746	1	62.173	1	62.078	1
Solar Module Vendor 3	55.704	3	57.222	2	56.715	2	56.547	3
Solar Module Vendor 4	48.738	4	48.621	4	48.645	4	48.668	4
Solar Module Vendor 5	47.099	5	46.855	5	46.005	5	46.653	5

$$BNP_{SM1} = \frac{(72.44 - 43.86) + (57.35 - 43.86)}{3} + 43.86 = 57.884$$

Similarly, other calculations can be carried out for the manufacturers, experts, and users decision-making groups. Table 9.15 lists the final ranking based on the BNP values of different decision-making groups.

9.8 CONCLUSION

The comparative BNP values of each of the decision-making groups as shown in Table 9.16 contribute some useful insights into the thought process. The maximum weightage by manufacturer group is ascribed to Financial Strength (0.131) and subsequently to Service Quality (0.108). For the experts group, the value is maximum for Service Quality (0.164) and then Financial Strength (0.096). For the users group, the BNP value is maximum for Service Quality (0.169) and next for Module Efficiency (0.103). All three decision-making groups are pointing to service quality, though it is considered to be the major factor for experts and users group. The manufacturers group considers financial strength as the major factor. The users group also desires module efficiency to be significantly considered in choosing solar panels.

The final ranking, as per different decision groups in Table 9.15, clearly shows Solar Module Vendor 2 as the best. However, Solar Module Vendors 1 and 3 are closely ranked at either second or third by different decision groups. This means that Solar Module Vendors 1 and 3 are similar in characteristics. On the other hand, Solar Module Vendors 4 and 5 lie at the bottom of the table. Different decision groups decide differently for sub-criteria, because their priorities are different. Users are more inclined towards better module efficiency and service quality, while

TABLE 9.16
BNP Value of Criteria and Sub-Criteria for Each Group and the Average Value

	Criteria & Sub-Criteria	Manufacturers	Experts	Users	Average
C1	**Performance Criteria**	0.127	0.199	0.304	0.210
C11	Module Efficiency	0.028	0.055	0.103	0.062
C12	Power Sorting	0.030	0.039	0.041	0.037
C13	Lifetime Energy Yield	0.028	0.038	0.055	0.040
C14	PTC Power Rating	0.025	0.040	0.049	0.038
C15	Warranty	0.016	0.027	0.056	0.033
C2	**Financial Properties**	0.226	0.074	0.067	0.122
C21	Government Support	0.077	0.035	0.035	0.049
C22	Price	0.078	0.020	0.012	0.037
C23	Cost per Watt	0.071	0.019	0.020	0.037
C3	**Technical Characteristics**	0.153	0.169	0.161	0.161
C31	Dimension	0.011	0.016	0.016	0.014
C32	Weight	0.013	0.015	0.023	0.017
C33	Durability	0.021	0.017	0.026	0.021
C34	Corrosion Resistance	0.016	0.021	0.013	0.016
C35	Fill Factor	0.027	0.026	0.016	0.023
C36	Recyclability	0.033	0.040	0.023	0.032
C37	Energy Payback	0.031	0.034	0.043	0.036
C4	**Customer Support**	0.167	0.228	0.230	0.208
C41	Service Quality	0.108	0.164	0.169	0.147
C42	Spare Parts Availability	0.059	0.064	0.061	0.061
C5	**Environment Friendliness**	0.089	0.121	0.091	0.100
C51	Area Covered	0.033	0.061	0.064	0.053
C52	Material	0.055	0.061	0.027	0.048
C6	**Supplier Reputation**	0.239	0.210	0.147	0.199
C61	Financial Strength	0.131	0.096	0.053	0.093
C62	ISO Certification	0.048	0.052	0.037	0.046
C63	Reliability	0.060	0.061	0.057	0.060

manufacturers tend towards financial conditions. The FAHP technique under Multi-Criteria Decision-Making (MCDM) thus provides an improved way of decision-making by harmonizing the decisions of different stakeholders.

REFERENCES

Buckley, J.J. (1985). Fuzzy hierarchical analysis. *Fuzzy Sets and Systems, 17*(3), 233–247.

Desideri, U., & Asdrubali, F. (2018). *Handbook of energy efficiency in buildings: A life cycle approach* (pp. 1–836). Butterworth-Heinemann, Oxford.

Hwang, C.L., & Yoon, K. (1981). Methods for multiple attribute decision making, Lecture Notes in Economics and Mathematical Systems, vol. 186 (pp. 58–191). Springer, Berlin, Heidelberg.

Metz, A., Fischer, M., & Trube, J. (2017 May). International technology roadmap for photovoltaics (ITRPV): Crystalline silicon technology-current status and outlook. In Proceedings of the PV Manufacturing in Europe Conference, Brussels, Belgium, pp. 18–19.

Jordan, D.C., & Kurtz, S.R. (2013). Photovoltaic degradation rates – An analytical review. *Progress in Photovoltaics: Research and Applications*, *21*(1), 12–29.

Jordan, D.C., Kurtz, S.R., VanSant, K., & Newmiller, J. (2016). Compendium of photovoltaic degradation rates. *Progress in Photovoltaics: Research and Applications*, *24*(7), 978–989. John Wiley and Sons Ltd.

Saaty, T.L. (1980) *The analytic hierarchy process*. McGraw-Hill, New York.

Sun, P., Yang, J., & Zhi, Y. (2019). Multi-attribute decision-making method based on Taylor expansion. *International Journal of Distributed Sensor Networks*, *15*(3), 1–9.

Takeda, E., Cogger, K.O., & Yu, P.L. (1987). Estimating criterion weights using eigenvectors: A comparative study. *European Journal of Operational Research*, *29*(3), 360–369.

Tzeng, G.H., & Huang, J.J. (2011). *Multiple attribute decision making: Methods and applications*. CRC Press.

Wang, T.C., & Tsai, S.Y. (2018). Solar panel supplier selection for the photovoltaic system design by using fuzzy Multi-Criteria Decision Making (MCDM) approaches. *Energies*, *11*(8), 1–22.

Webber, J., & Riley, E. (2013). Mismatch loss reduction in photovoltaic arrays as a result of sorting photovoltaic modules by max-power parameters. *ISRN Renewable Energy*, *2013*, 1–9.

Weckend, S., Wade, A., & Heath, G. (2016). *End of life management solar PV panels*. www.irena.org.

What Does Module Efficiency Mean? (December 2016). Retrieved in September 2020 from https://www.infiniteenergy.com.au/module-efficiency-mean/.

Xu, J., & Wu, W. (2006). Multiple attribute decision making theory and methods. Tsinghua University Press, Beijing, China.

10 Using Multi-Criteria Decision-Analysis to Build a Long-Term Business Strategy

Sandeep Bhasin

CONTENTS

10.1 Introduction 179
10.2 Connecting MCDA with Strategic Components 181
10.3 Strategic Decision-Making through Scenario Planning 182
10.4 The Proposed Two-Step Process of Validating a Strategy 185
10.5 Systems Thinking and Strategic Decision-Making 186
10.6 Conclusion 187
Note 187
References 188

10.1 INTRODUCTION

Building a sustainable business model has become a real challenge even for established market leaders. A study of a Fortune 500 ranking shows that there are a few organisations that performed consistently since the listing was first published in 1955. Out of these organisations, only one organisation has managed to stay on the list over the decades. This fact reveals a real problem faced by any organisation: Building a sustainable and relevant business model which can aid the organisation survive the business cycles over a period of time. This can be achieved by conceptualizing a strategy that makes the organisation stay relevant and is also useful in beating the direct and indirect competition. In this study, we will look at the efficient use of MCDA, which can help the organisation build a sustainable business model.

Monopolies, in a real sense, do not exist, as there is always a substitute available for consumers. With active competition, direct and indirect, businesses across industries have to create and maintain their positioning by using a plan that can work as a deterrent to the existing and new competition. These long-term plans to achieve the mission and vision of the organisation are called strategies. Organisational strategy is the creation of a unique and value-based offering involving a different set of activities (Porter, 1996) than the competition. If there were only one ideal direction for a business to work for, then a strategy would be unnecessary. These strategies are the relationships among the ecosystem-driven structures of an organisation and the behaviour of different stakeholders. Every strategy is dependent on 'systems' (Meadows, 2008) which they build over a period of time.

Those organisations that build a unique system that produces acceptable results over a period of time has proven to be sustainable in their approach (Jahan et al., 2016). Most of these strategic decisions are based on extensive research done at the grassroots level, supported by a better understanding of macro and micro economic factors. A number of these well-thought-through strategies fail, and the impact of these failures is devastating.

If we build a system to help organisations opt for a well-planned strategic call, the failure rate may drastically reduce. With an MCDA approach, organisations can reduce the probability of failures. Broadly, MCDA is similar to a cost-benefit analysis but with the notable advantage of not being solely limited to monetary units for its comparisons (Janse, 2018). By systemizing complex issues and analyzing multiple scenarios of outcomes, an organisation can guide itself to better informed, justified decisions that, in the long run, can help the corporation build a better, sustainable business model Smith and Weistroffer (2000). Traditionally, MCDA problems are comprised of five components (Natural Research Leadership Institution – NRLI 2011):

- The goal (could be short-term, mid-term or long-term).
- Decision-makers with opinions and preferences.
- Decision alternatives.
- Evaluation criteria (interests).
- Outcomes or consequences associated with alternative interest combination.

These five components can be shrunk to understand the criteria, projecting the alternatives, and weighing the attributes and performance ratings of available alternatives (Mousavi-Nasab & Sotoudeh-Anvari, 2017). Every strategic decision considered by the corporates follows a similar path, but the dynamism changes with the addition of a new element compared to the competition, direct or indirect. With the inclusion of competition as one of the factors, building scenarios as a part of the MCDA process can help the corporate assign probabilities to all the possible outcomes, thus connecting the concept of MCDA with the business strategic building of an organisation.

The five components suggested above describe the perspective of MCDA in operations-related decisions, but we can extend the same approach into a highly complex business strategy domain.

For any strategy to be effective, it must pass the five following tests:

a. *Purpose:* A change in purpose changes a system profoundly, even if every element and interconnection remains the same.
b. *Macro Environment Analysis:* Every organisation operates in a highly dynamic macro environment ecosystem. This ecosystem is effectively controlled by the stakeholders, including the competition and the government.

A good understanding of the ecosystem helps an organisation in creating a strategy while keeping the purpose in mind.

c. *Micro Environment Analysis:* The organisation creates its own ecosystem based on macro analysis, defining the overall performance in line with this macro analysis.
d. *Competition:* Competition is not just from the direct players who sell similar products in the market space. Competition can also come from substitutes, which can disrupt an industry. A perfect understanding of the competition defines the trajectory of growth for the organisation (Kumar et al., 2017).
e. *Sustainability:* For an organisation to survive, it has to not just operate on a sustainable business model, but also ensure its business model to be resilient, flexible, and highly adaptable (Mittal et al., 2018; 2019).

The business decisions considered by corporates at the strategic level arc supported by data from the past which may consequently not help in the future because of the dynamics followed by the markets. The following section attempts to connect the MCDA concept with the components of strategy at various levels, which the decision-maker may use to work on for better synergy between the operations-driven strategy and the long-term business strategy.

10.2 CONNECTING MCDA WITH STRATEGIC COMPONENTS

The generally accepted components of MCDA, restricted predominantly to the short-to-mid-term operations strategy are stretched to the strategic level, as shown in Figure 10.1. The component of the goal can be aligned with the strategic purpose of an organisation. This is well-defined and publicized within the organisations by the management, especially to attain buy-ins from internal customers working on the value-chain. It may be noted that any purpose without a buy-in from the internal customers has a relatively low probability of success (Sindhwani et al., 2018). The other four components of MCDA, viz., the opinion and understanding of the decision-makers, the decision options, or available alternatives with the decision-makers based on set criteria with a definitive outcome, are directly related to the micro-environment and macro-environment in which the organisation operates. These components collectively can be paralleled to the strategic components of Macro Economic analysis, Micro Economic analysis, and short-term and long-term impact of these factors on competition behaviour and the conclusive impact on business model's sustainability. The overlap of these activities draws up a footing for extending the components of MCDA to the overall strategic outlook of an organisation.

For any strategy to become a success, it needs to align itself with the corporate's mission and the promoters' vision (Purpose) followed by a complete analysis of the Macro, Micro, and Meso levels of the ecosystem in which the corporate works. Any dissonance in the understanding of the strategic outcome can lead to a higher probability of failure.

As a decision-maker, one has to manage these risks of failure. From an organisation's perspective, these risks can be categorized as Level-One risks, which, if not addressed, can result in failure of the business itself; Level-Two risks, which

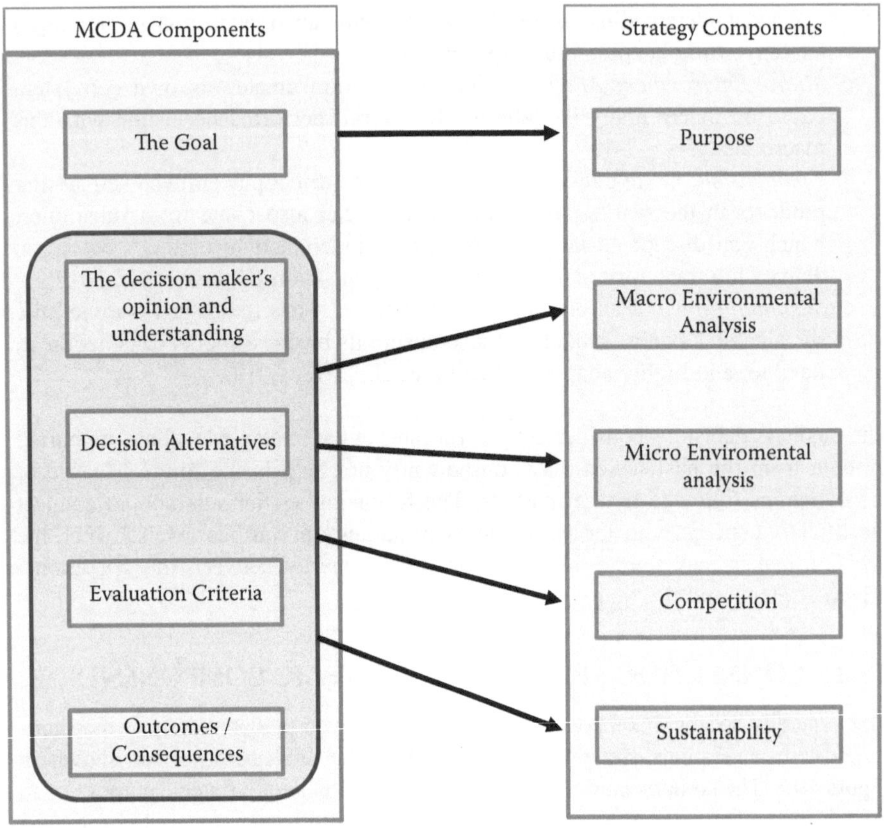

FIGURE 10.1 Linking MCDA and Strategy Components.

can impact the next couple of years but may not impact the organisation's future; and Level-Three risks, which corporates face all of the time. The success of any strategy depends on the alignment of the purpose with the business model, connected with resilience and sustainability (Bhasin, 2019). Using the concept of MCDA, we can design a process that may help the organisation define its long-term strategy. MCDA has been traditionally used to address Level-Two and Level-Three risks; however, we may use the same principles in addressing Level-One risks as well. Even though numerous theoretical perspectives are available to address the decision-making process, we utilize the (strategic) Scenario Planning concept to define the flow of decision-making using MCDA, as presented in Figure 10.1 above, in the next section.

10.3 STRATEGIC DECISION-MAKING THROUGH SCENARIO PLANNING

Scenario Planning addresses the outcome-related issues, as every action implemented by the decision-maker would lead towards multiple possibilities based on relative weights assigned while considering the macro, micro, and meso environmental analysis, as addressed above.

The environment (external and internal, Macro, and Micro) in which an organisation operates can change the outcome by a great margin. If we add the possibility of a Black Swan event[1] and the probability of the desired or expected outcome for the corporate, then this becomes even more disturbed.

Scenario Planning implicates continuous learning from past records in order to help predict the response of a specific decision in the future (De Geus, 1988). The calculation of probability helps in the construction of possible scenarios. The organisation's vision and mission must be valid enough from the long-term perspective in order to work. Once the board of the organisation finalizes the mission and the vision, the following step is to build the scenarios and plan the strategy to achieve these predefined long-term goals. These goals must be dynamic enough to adapt to the changes forced upon by the micro and macroeconomic factors.

Whether we examine the components of MCDA or of strategic-planning, we heavily rely on the understanding of the decision-makers of the given situation. These understandings guide the organisations to their strategic goals. The opinion of the decision-makers and their understanding are based on the learnings of the past, but there also has to be an element of understanding of the future. For any strategist, this understanding needs to be validated in the market space. These are validated using qualitative and quantitative surveys using statistical tools. These studies offer the decision-makers a list of probable outcomes, based on which the decisions are taken. It may be worthwhile to note that these decisions must be considered complying with the categorization of the Three-Level risk parametre. Many organisations have effectively used the Three-Level risk parametre to help survive business risks by using the concept of scenario planning and probability. When a corporate faces the liability of choosing the link of business, such as in the case of Walmart in India, where they had to choose the area of business (i.e. B2B and B2C), they worked closely with their partners Bharti Enterprises in India and developed their presence in both of the segments with Cash-and-Carry for wholesale and EasyDay for retail. After taking over Flipkart, they have also added the scenario of using the brand in wholesale with Flipkart-Wholesale, which has a more efficient reach as well as brand acceptability in the country. Using the components of MCDA, we can draw upon the following relations:

In the micro-environment and macro-environment which Walmart has been operating, the consequences of the outcome would completely depend on national policies which, for a long time, restricted them from progressing into retail space. After taking over Flipkart, they have managed to get into retail space; However, in spite of that, they can only offer a platform to help traders meet the consumers. Building scenarios and assigning probabilities to the outcomes helped them take the decisions of not just buying existing but unrelated businesses, but also considering decisions of parting ways with partners. As presented in Figure 10.3, Walmart attained a better understanding of the impact of government policy decisions at the Macro level, the strategies of the competition at the Meso level, and subsequently recreated its strategic decisions at the Micro level in order to draw up a long-term strategic plan to be implemented for a sustainable business. The scenarios created accordingly, as illustrated in Figure 10.2 and Figure 10.3, need to be validated by the strategy. To corroborate the strategies, we use the risks as one of the main parametres, as outlined in Table 10.1.

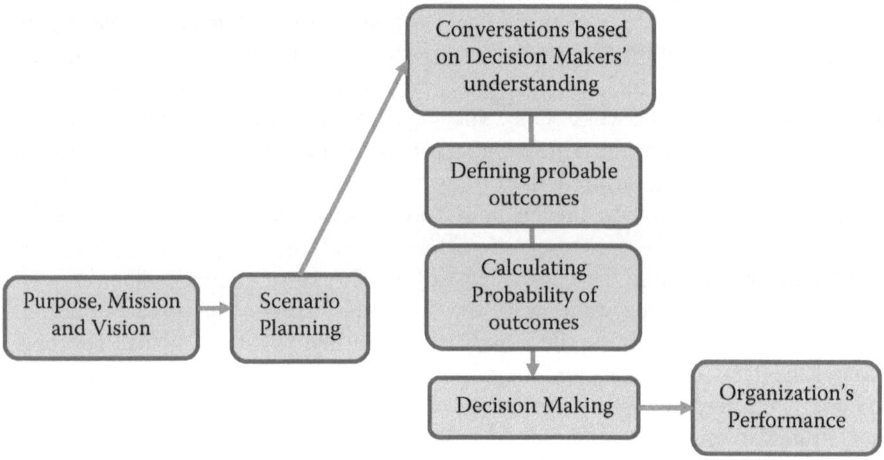

FIGURE 10.2 Scenario Planning Framework (Adaptation – Chermack, 2011).

FIGURE 10.3 Connecting Scenario Planning with MCDA.

TABLE 10.1
Three-Dimensional Analysis – Risk, Probability of Failure, and the Impact of Decisions

Probability	Risk			Impact
	Level-One	Level-Two	Level-Three	
High	0	0.5	1	Low
Medium	0.5	1	0	Medium
Low	1	0	0.5	High

10.4 THE PROPOSED TWO-STEP PROCESS OF VALIDATING A STRATEGY

The success of a strategy depends on factors that may not be in direct control of the decision-makers. For any strategist, capturing the essence of the 'chance' of succeeding in the opportunity is main focused area.

Step-1: Taking into account all of the dynamism of the market space, including direct competition and indirect competition from the substitutes, we can obtain the strategic alignment with the formula:

$$(S)^t = \sum_{k=0}^{n} (ks - n)x^t a^{sn} \tag{10.1}$$

Where,

S = Strategy

s = Strategic input of the competitor/substitute which is better or worse than the self, based on a 10-point scale

t = Time horizon in years

k = Number of competitors, as defined by the decision-makers

n = Expected number of competitors in 't' years

x = Constant, based on macro data points

a = Number of substitutes

The objective of strategy (S) over a period of time (t) is to measure the impact of competition, including the substitutes available in the market space, while constantly maintaining the new competition and the substitutes on the horizon. All of these units must be validated using the data collected from the market. After completing a proper analysis of the impact of strategy at any given point in time, we expand our understanding of the outcome by conducting the next level of analysis by adding the dynamism of the market by calculating the probability of a particular outcome.

Step-2: The strategic alignment of the corporate with the purpose, mission, and vision can be calculated if the decision-makers can build multiple scenarios. The major element is as follows:

$$f(S) = a_0 + \sum_{k=1}^{n+1} \left(a_n xs \frac{ntx}{L} + b_n xs \frac{ntx}{L} \right) \qquad (10.2)$$

S = Strategy
a = Number of competitors
b = Number of substitutes
t = Time horizon in years
k = Increase in the number of competitors
n = Expected number of competitors/substitutes in 't' years
x = Constant
L = Gestation period for the industry

The above equation can help the organisation validate the functionality of Step One by adding the element of a *gestation period* in the calculations, taking into consideration all of the aspects that impact the outcome of the strategy in the mid-to-long term. By applying these two steps, the decision-makers in an organisation can reduce the probability of failure by making well-informed decisions while taking into consideration various scenarios of outcomes. By calculating the probability with the associated risks, the decision-maker is expected to accommodate the decisions on a three-dimensional table as shown below:

Step-3: Level-One risks are high-risk decisions that can make or break the business model. The impact of such risks will be significant if not addressed well in time, even if the probability of occurrence is low. As discussed earlier, the Level-One risks should be examined before any other risks. Level-Two risks are medium-level risks that offer time for the decision-makers to be able to rectify a path that they can take in order to solve the issues. Such risks will have a medium impact on the business even as the probability of occurrence is medium. Level-Three risks are the day-to-day risks that every business faces. Even if the probability of the occurrence of such risks is high, the impact remains low. Any strategic call conducted by the corporates has a long-term impact. This can not only be better drawn using Systems Thinking, but can also help the corporates in building a process flow of activities which is in sync with the scenarios.

10.5 SYSTEMS THINKING AND STRATEGIC DECISION-MAKING

The system structure is the source of system behaviour. System behaviour, in turn, reveals itself as a series of events over time. As an organisation plans to address the latter two categories of risks, systems thinking can be used to measure the efficiency and validity of the processes at length. As represented in Figure 10.3, every input can be controlled before the process of analyzing the input starts. Figure 10.4 and Figure 10.5 further presents this procedure.

The same holds true for the output, as the organisation can control the output by managing the process. The dynamism increases when outsourcing is added to the process. Many organisations opt for outsourcing the process in order to save costs. They outsource when they do not possess the expertise of the process, and this is where the third-party vendors come in. With the entry of third-party vendors, the dynamism of the process changes because now, the third-party vendors can work closely with the substitutes and the competition at the same time, thus resulting in a

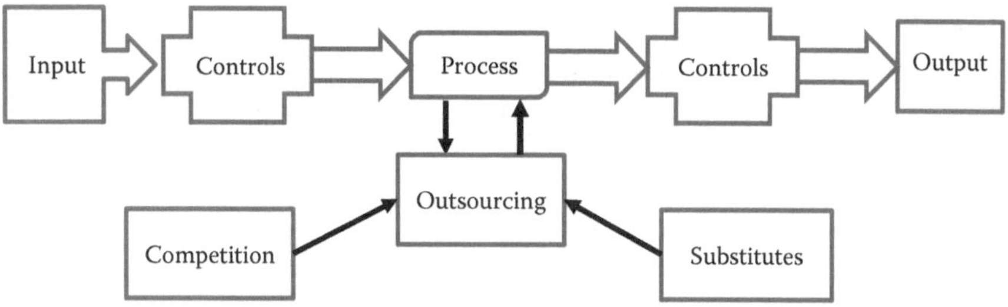

FIGURE 10.4 Systems Thinking Flow.

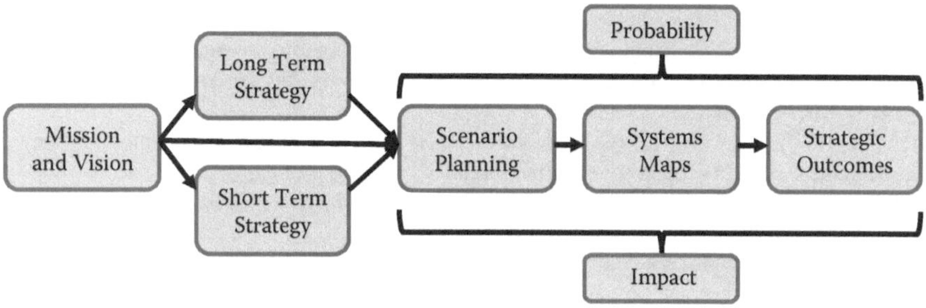

FIGURE 10.5 The Flow.

dent in the organisation's mid-term and long-term strategies. The systems process, combined with scenario planning, helps in building the decision-process that works towards achieving the long-term goals of an organisation.

10.6 CONCLUSION

The Scenario Planning and Systems Thinking processes, together with the basic principles of Multi-Criteria Decision-Making, help the planners in using both intuitive powers as well as rational decision-making processes to be able to achieve the ultimate goal of decisions that are aligned with the vision and the mission of the organisation. The complete process starts with the definition of the vision and the mission of the organisation, which gets broken into long-term and short-term strategies. These well-defined strategies can be achieved by building scenarios using specific techniques. These scenarios are then followed by defined steps for building a systems map for every activity in order to acquire a controlled strategic outcome. The methods mentioned in the chapter have been used individually by corporates. Connecting these methods using mathematical formulas and both rational and intuitive processes can help the organisation in building a well-defined system within the process followed.

NOTE

1 A Black Swan Event, as described by Nassim Nicholas Taleb, must have the following three characteristics: The event should be an outlier; the event must have an extreme impact; and it should not be predictable.

REFERENCES

Bhasin, S. (2019). Common pool resource – The problem of sustainability. *International Journal of Emerging Technologies and Innovative Research, 6*(3), 228–238.

Chermack, T.J. (2011). *Scenario planning in organization* (pp. 35). Berrett–Koehler Publishers Inc.

De Geus, A. (1988). Planning as learning. *Harvard Business Review, 66*(2), 70–74.

Jahan, A., Edwards, K.L., & Bahraminasab, M. (2016). Multi-criteria decision analysis for supporting the selection of engineering materials in product design, second edition (pp. 1–52). Butterworth-Heinemann, Oxford, https://doi.org/10.1016/C2014-0-03347-3

Janse, B. (2018). Multiple Criteria Decision Analysis (MCDA). ToolsHero. Retrieved [insert date] from https://www.toolshero.com/decision-making/multiple-criteria-decision-analysis-mcda/

Kumar, R., Kumar, V., & Singh, S. (2017). Work culture enablers: Hierarchical design for effectiveness & efficiency. *International Journal of Lean Enterprise Research (IJLER), 2*(3), 189–201.

Porter, M. (1996). What is strategy? *Harvard Business Review*, November–December. Strategic Management of Technology and Innovation.

Meadows, D.H. (2008). *Thinking in Systems: A Primer. White River Junction*. Chelsea Green Publishing, USA.

Mittal, V.K., Sindhwani, R., Shekhar, H., & Singh, P.L. (2019). Fuzzy AHP model for challenges to thermal power plant establishment in India. *International Journal of Operational Research, 34*(4), 562–581.

Mittal, V.K., Sindhwani, R., Singh, P.L., Kalsariya, V., & Salroo, F. (2018). Evaluating significance of green manufacturing enablers using MOORA method for Indian manufacturing sector. In S. Singh, P. Raj, & S. Tambe (Eds.), *Proceedings of the international conference on modern research in aerospace engineering*. (Lecture Notes in Mechanical Engineering) (pp. 303–314). Springer, Singapore.

Mousavi-Nasab, S.H., & Sotoudeh-Anvari, A. (2017). A comprehensive MCDM-based approach using TOPSIS, COPRAS and DEA as an auxiliary tool for material selection problems. *Materials and Design, 121,* 237–253.

Natural Research Leadership Institution – NRLI. (2011). https://tinyurl.com/y68yhsxk

Sindhwani, R., Mittal, V.K., Singh, P.L., Kalsariya, V., & Salroo, F. (2018). Modelling and analysis of energy efficiency drivers by fuzzy ISM and fuzzy MICMAC approach. *International Journal of Productivity and Quality Management, 25*(2), 225–244.

Smith C.H., & Weistroffer, H.R. (2000). On designing health care plans and systems from the multiple-criteria decision making (MCDM) perspective. In Y. Y., Haimes, & R. E., Steuer (Eds.) *Research and practice in multiple criteria decision making*. (Lecture Notes in Economics and Mathematical Systems, vol. 487). Springer, Berlin, Heidelberg.

11 Systematic Review and Deliberation of Various Multi-Criteria Decision-Making Techniques

Muskan Jindal and Areeba Kazim

CONTENTS

11.1 Introduction ... 189
11.2 Analytical Hierarchy Process (AHP) ... 191
11.2.1 Gradation 1 Inaugurate Gradation for the "Criteria and Alternatives" ... 191
11.2.2 Gradation 2 Create a Couple-Wise Desertion Decision Matrix (M) ... 192
11.2.3 Gradation 3 The Matrix is Normalized and Prioritized to Get the Respective Weights of Criteria ... 193
11.2.4 Gradation 4 Evaluate the Reliability of the Judgements to Confirm the Outcome ... 193
11.3 Analytical Network Process (ANP) ... 195
11.4 Technique for Order Preference by Similarity to an Ideal Solution (TOPSIS) ... 196
11.5 Vlse Kriterijumska Optimizcija I Kaompromisno Resenje (VIKOR) ... 198
11.6 ELimination Et Choix Traduisant la REalité (ELECTRE) ... 198
11.7 Fuzzy Logic ... 199
11.8 Conclusion ... 200
References ... 200

11.1 INTRODUCTION

Multi-Criteria Decision-Making (MCDM) is a methodology that can be defined as a technique used in dealing with pivotal judgement, thereby managing complex tasks that have variegated criteria and multifarious scenarios and attributes to deliberate (Kumar et al., 2017; Shanker et al., 2019; Sindhwani et al., 2018; Sindhwani & Malhotra, 2017; Sindhwani, Singh, Iqbal, et al., 2019). Often in complex decision-making problems, there are multiple objectives, diverse attributes to consider, and various boundaries and values to abide by (Kumar et al., 2019;

Sindhwani, Singh, Chopra, et al., 2019). Moreover, due to a number of attributes, many decision-making tasks have predefined priorities, thus making the task at hand more intricate and variegated. MCDM was first introduced by Pomerol and Romero (Barba-Romero & Pomerol, 2000), when they deliberated various scenarios and attributes in order to obtain a feasible choice.

This process of multi-criteria decision-making works in a graded and structural manner by primarily describing the objectives and their respective significance and comprehending the criteria and alternatives along with their individual attributes, if any. The subsequent gradation is done to quantify relevant information by transforming pertinent attributes and criteria to their respective quantifiable units. Next, the succeeding gradation is conducted to deliberate various attributes to be able to provide them with their priority weights. The final step is the application of various mathematical algorithms and calculation models (Pohekar & Ramachandran, 2004). Real-life decision-making problems are abstruse and convoluted in nature due to the presence of unknown, labyrinthine, and sometimes contradictory attributes or objectives. Moreover, most of the deliberated problem statements have multifarious and multiple criteria which may or may not complement or contradict each the other, such that all the criteria attributed are unique and bear a different set of priorities. Thus, most real-life scenarios require and utilize MCDM to comprehend and provide optimal solutions, as these complex and challenging problems are usually in the realm of human understanding, knowledge, and attainment (Kumar et al., 2017a). Thus, MCDM techniques and toolkits allow judgement makers to apply a scientific approach and explore various attributes and avenues before making the final judgement, thereby decreasing time duration while increasing efficiency and simultaneously reducing the risk quotient (Mittal et al., 2018).

In the explored, yet niche, domain of MCDM, the concept of ambiguity, uncertainty, and incomplete information is quite mundane yet challenging to conquer for any MCDM practitioner. This includes situations and case studies where the information, data, attributes, tables, description, and tables provided are insufficiently detailed, bear missing attributes and values, or may or may not carry cogent qualitative data to support or qualitative information to cogent the case's understanding or predict, prove, or develop any model to so (Zimmermann, 2000). This has been rather deftly and skillfully handled in the field of MCDM, as toolkits and the existence of a plethora of methods, terminologies, algorithms, and models are available in the currently published literature. Since uncertainty can be of variegated types and of different levels depending on the amount of information, it is vital to consider the priority of ambiguous information, the methodology used, the user's comfortability, and the tools utilized to solve these mathematical equations (Saaty, 1980). However, the selection of optimal methods is completely dependent on attributes, scenarios, and methodology implemented, and it is salient that not all MCDM approaches are suitable. In fact, every approach has a different set of algorithms that produce the best results, so many times, researchers prefer hybrid methods executed by amalgamating various optimal methods to attain the best results that cater to the set of complexities raised due to missing information and subsequently fed to a disparate set of individual problems unique to their model and case.

11.2 ANALYTICAL HIERARCHY PROCESS (AHP)

The Analytical Hierarchy Process was created in 1980. As one of the most generally utilized MDCM approaches, the Analytical Hierarchy Process is equipped for helping model choice, standards significance examination, and elective assessment (Mittal et al., 2019). The procedure can be expected to function when the subjective and quantitative parts of judgement are incorporated as well (Vaidya & Kumar, 2006). The Analytical Hierarchy Process utilizes the idea of couple-wise correlations with improving the effectiveness of combining qualitative and quantitative assessments in a judgement procedure. It contains various other options and models for making a decision about other options. The methodology permits judgement creators to communicate their sentiments by contrasting two options one after another, as opposed to simultaneously with all of the other options. It rearranges and facilitates a decision-making process on abstruse problem sets.

The observability and effectiveness attributes of the Analytical Hierarchy Process add to its ubiquity across various enterprises. Certain references (Wang et al., 2015) revealed that the Analytical Hierarchy Process method has been utilized in about 150 applications. Models utilizing the Analytical Hierarchy Process in delivery and sea segments incorporate port judgement and competitiveness evaluation (Yeo et al., 2010), vessel choice (Carlos Perez-Mesa et al., 2012), port allocation (Ung et al., 2006), hazard estimation of boat activities (Sii & Wang, 2003), plan bolster evaluation for the seaward business (Ugboma et al., 2004), port help quality positioning (Akyuz et al., 2015), sea guideline usage (Beşikçi et al., 2016), transport operational vitality effectiveness (Chou & Ding, 2016), judgement of boat banner (Akyuz et al., 2015), evaluation of the sea work showing consistency (Sahin & Senol, 2015), and marine mishap analysis (Chen & Lin, 2006). The Analytical Hierarchy Process utilizes a mathematical procedure to handle emotional judgements of an individual or the gathering in a judgement making process. It is comprised of four stages:

1. Building a progressive system of characteristics and substitutes.
2. Making couple-wise examinations for quality and evaluating loads of the standards and the comparative execution standards designated to substitutes concerning every rule.
3. Accumulating the loads and execution values for alternative needs.
4. Examining the reliability of the judgements to check the outcome produced.

11.2.1 Gradation 1 Inaugurate Gradation for the "Criteria and Alternatives"

The pyramid chain is a fundamental concept of the Analytical Hierarchy Process. In order to lead an AHP study, a chain of the hierarchy of clear rules and choices should be built. Figure 11.1 displays a case of progression with characterized models and options.

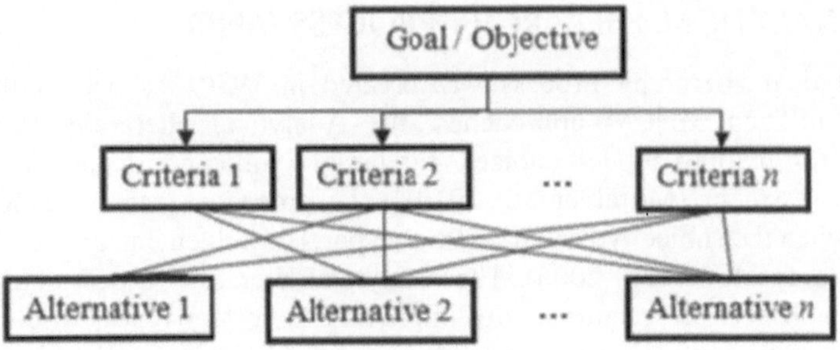

FIGURE 11.1 Hierarchy Chart for Criteria and Alternatives.

11.2.2 Gradation 2 Create a Couple-Wise Desertion Decision Matrix (M)

A couple-wise contrast matrix (M) is developed catering to all existing attributes. Eq. 11.1 embodies an enumerated conclusion on a couple of criteria that represent the significance of 'Criteria 1(C1)' over 'Criteria 2 (C2).' A gauge of 1–9 is implemented to pursue non-enumerated couple-wise assessments of the two deliberated elements. Finally, Table 11.1 refers to the verbal judgements. Figure 11.2

$$M = \begin{array}{c} C_1 \\ C_2 \\ \vdots \\ C_j \\ \vdots \\ C_n \end{array} \overset{\begin{array}{cccccc} C_1 & C_2 & \cdots & C_i & \cdots & C_n \end{array}}{\begin{bmatrix} a_{11} & a_{21} & \cdots & a_{i1} & \cdots & a_{n1} \\ a_{12} & a_{22} & \cdots & a_{i2} & \cdots & a_{n2} \\ \vdots & \vdots & & \vdots & & \vdots \\ a_{1j} & a_{2j} & \cdots & a_{ij} & \cdots & a_{nj} \\ \vdots & \vdots & & \vdots & & \vdots \\ a_{1n} & a_{2n} & \cdots & a_{in} & \cdots & a_{nn} \end{bmatrix}} \quad (11.1)$$

TABLE 11.1
Conclusion Marks in AHP

Marks	Conclusion	Description
1	Similar	Two exercises contribute similarly to the goal.
3	Reasonably	Experience and judgement somewhat favour one movement over another.
5	Adamantly	Experience and judgement firmly favour one movement over another.
7	Very Adamantly	An action is unequivocally preferred, and its predominance showed while practically speaking.
9	Extremely	The proof preferring one movement over another is the most noteworthy and conceivable request of insistence.

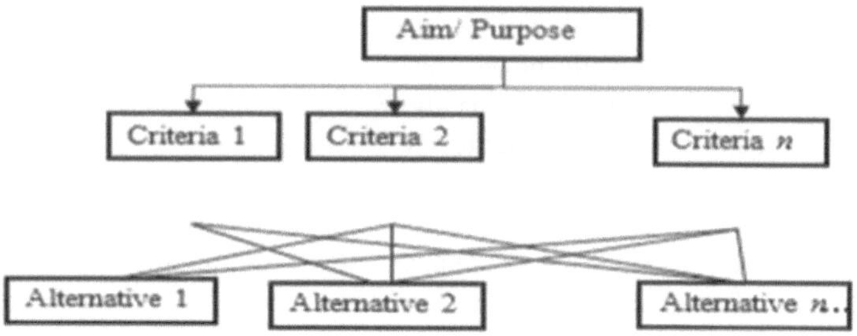

FIGURE 11.2 Chain of Command Hierarchy for Criteria and Alternatives.

11.2.3 Gradation 3 The Matrix is Normalized and Prioritized to Get the Respective Weights of Criteria

The purpose of normalizing the matrix is to calculate the respective weight of each deliberated criteria. To implement the same aggregate of each vertical column, the value is achieved by performing the addition of respective values. In other words, all of the respective values of kth horizontal rows in the matrix are computed by their aggregate mean to obtain the kth criteria's aggregate weight, as mention in Equation (2.2):

$$w_k = \frac{1}{n} \sum_{j=1}^{n} \frac{a_{kj}}{\sum_{i=1}^{n} a_{ij}}, \quad k = (1, 2, \ldots n) \tag{11.2}$$

Where, a_{ij} represents the section of line 'I' and segment 'j' in a correlation lattice of request 'n,' and 'w_k' is the weight of a particular rule 'k' in the pairwise examination matrix.

11.2.4 Gradation 4 Evaluate the Reliability of the Judgements to Confirm the Outcome

In order to determine important weights, insignificant reliability is required and a test must be accomplished. The reliability of the examination matrices is followed by the 'Consistency Ratio (CR).' The CR Index in the Analytical Hierarchy Process is utilized so as to maintain reliability in dynamic outcomes.

$$CR = \frac{CI}{RI} \tag{11.3}$$

CI represents the reliability index, and RI represents the arbitrary index.

This method was applied by Kagazyo (Kagazyo et al., 1997), where resource, social criteria, and technological issues were assessed in prioritization of energy projects to identify that the electricity supply system in Japan consisted of the maximum amount of power generation from combustion of coal. The AHP method was used by Akash (Akash et al., 1999) to compare various energy generation alternatives and to conclude that the best substitutes were hydropower energy, solar energy, and wind energy, while nuclear energy and fossil fuels proved to be the worst for electricity generation. Xiaohua and Zhenmin (Xiaohua & Zhenmin, 2002) developed an index system to evaluate the relationship between sustainable development and energy in remote locations, using AHP to calculate the weight of each index. Furthermore, this was also applied at a university campus in Eskeşehir, Turkey by Aras (Aras et al., 2004) to determine the most suitable site for placing a station for wind observation, where the most significant criteria selected were security and the regional topography. A methodology for making decisions on the basis of the AHP process was proposed by Kablan (Kablan, 2004) to prioritize the policy tools related to the conservation of energy from a Jordan-based case study. It was observed that for the conservation of energy resources, the most applicable policies indicated were namely, 'law-making and execution,' 'fiscal incentives,' and 'qualification, education, and training.' A methodology was developed by Wang (Wang et al., 2008) on the assessment of grey incidence for distributed systems on triple generation, which involved procedures of entropy information and AHP along with the weighting of linear combinations. Five alternatives were evaluated using this model, thus obtaining distinct results for each assessment.

Talaei (Talaei et al., 2014) used this method in Iran to rank various novel technologies related to low-carbon energy based on various criteria. They concluded that oil and gas have the highest priority in technological usage for electricity and transport. A decision-making framework was further improved upon by Rosso (Rosso et al., 2014) with the application of the AHP method in providing an effective assessment of hydropower projects for the construction of hydropower plants in mountain areas. They integrated multi-criteria evaluation and stakeholder analysis to determine that the first project is the most applicable alternative. A model was proposed by Bojesen (Bojesen et al., 2015) for planning and making decisions in which they applied multi-criteria estimation from a spatial perspective and AHP to calculate priorities of the criteria, thereby defining and ranking the locations in Denmark that are convenient for the production of biogas. AHP was also applied (Al Garni et al., 2016) to define an evaluation criterion for assessing wind turbines in order to develop a wind farm. Four wind turbines were prioritized, and the best option that was selected was T-1. The method was again used to estimate the suitability of wind port alternatives for the three lifecycle phases of wind projects positioned offshore (Shirgholami et al., 2016). Two main important criteria were chosen, and the Sheerness port was considered as the best alternative at the end of the research study. The deployment of solar energy in the Indian region was estimated (Akbari et al., 2017) by conducting a Strength-Weakness-Opportunities-Challenges (SWOC) analysis, prioritizing the variables by utilizing the AHP technique. In Paraguay, four energy policy alternatives were prioritized and evaluated (Sindhu et al., 2017) using an MCDM model based on AHP under five main criteria to be able to determine that the most suitable course of action was the development of industrial zones of limited size. An integration of the MCDM model and AHP was proposed (Blanco et al., 2017) for renewable energy resources in Algeria to be evaluated and the alternatives to be ranked accordingly. In Jordan, different scenarios were developed, and the alternatives of energy resources were ranked correspondingly by (Haddad et al., 2017) using AHP. Their results conclude the conventional sources as the more suitable alternative for the area. Finally, (Malkawi et al., 2017) evaluated six alternatives with various main criteria and sub-criteria to rank the best wind turbine.

11.3 ANALYTICAL NETWORK PROCESS (ANP)

The Analytical Hierarchy Process has a few points of interest over different strategies in light of its effortlessness and its capacity to rank pieces of a multi-criteria issue in a various levelled structure (Sagbansua & Balo, 2017). However, it comes up short on the capacity to display the interdependencies among the criteria, which obliges its applications in complex frameworks, for example, in transport networks. The Analytical Network Process (ANP) (Saaty, 1990) was created to supplement AHP such that the criteria are introduced in a network, rather than the order, structure. ANP, being appropriate for presenting interdependency among the choice components, becomes a helpful MCDM instrument. It is an augmentation of AHP and permits the consideration of association among and between levels of criteria and choices. ANP utilizes a network without the need to indicate the levels in a pecking order. It provides a sensible method of managing reliance. Networks in ANP incorporate a number of components that may impact one another.

A pairwise examination matrix is set up for all of the components in Equation 11.4. The respondents need to address the inquiries. An example is 'Given a component and its upper-level target, which of the two components impacts the given component more with respect to the upper-level goal, and how much more is the impact compared to another component?' The reactions are introduced numerically, scaled based on Saaty's 1–9 scale, as seen in Table 11.1, where 1 represents a lack of interest between the two components, and 9 symbolizes the overwhelming strength of the component viable (in thc line of the matrix) over the examination component (in the segment of the matrix). The neighbourhood weights for all of the components are then created by utilizing Equation 11.5. The nearby weights attained from the pairwise correlation matrices become a piece of the contributions of a supermatrix.

$$A = a_{(ij)} = \begin{matrix} & \begin{matrix} e_1 & e_2 & \cdots & e_j & \cdots & e_m \end{matrix} \\ \begin{matrix} e_1 \\ e_2 \\ \vdots \\ e_i \\ \vdots \\ e_m \end{matrix} & \begin{bmatrix} a_{11} & a_{12} & \cdots & a_{1j} & \cdots & a_{1m} \\ a_{21} & a_{22} & \cdots & a_{2j} & \cdots & a_{2m} \\ \vdots & \vdots & \vdots & \vdots & \vdots & \vdots \\ a_{i1} & a_{i2} & \cdots & a_{ij} & \cdots & a_{im} \\ \vdots & \vdots & \vdots & \vdots & \ddots & \vdots \\ a_{m1} & a_{m2} & \cdots & a_{mj} & \cdots & a_{mm} \end{bmatrix} \end{matrix} \tag{11.4}$$

Assume that there are m components to be considered in a matrix, let $e_1, e_2, \ldots, e_m$ mean the various components, where $a_{(ij)}$ is the degree of impacts that the respondents conceptualize when component e_i is contrasted with e_j. When scoring is directed for a couple, the proportional worth is naturally allocated on the opposite correlation inside the matrix.

$$w_k = \frac{1}{m}\sum_{j=1}^{m}\frac{a_{kj}}{\sum_{i=1}^{n} a_{ij}} \quad k = (1, 2, \ldots m) \tag{11.5}$$

Where, w_k is the triage or prioritized vector of the kth component in the couple-wise examination matrix.

A network structure and a pairwise examination relationship are developed inside the structure. It takes into consideration reliance and incorporates freedom. It can organize gatherings or groups of components. It can deal with relationships better than AHP and 'can bolster a complex, networked dynamic with different impalpable criteria' (Tsai et al., 2010). In any case, ANP has two impediments: Right off the bat, it is hard for specialists to give the right network structure among criteria in any event, and various structures lead to various outcomes. Moreover, in order to frame a supermatrix, all criteria must be couple-wise contrasted with respect to every single criterion, which is troublesome and furthermore, unconventional (Yu & Tzeng, 2006).

For the evaluation of energy options in the region of Turkey, Ulutas (Ulutas, 2005) used the ANP technique to determine that the most suitable alternative of energy resource is biomass. (Erdogmus et al., 2006) evaluated fuel alternatives for the use of residential heating by applying the ANP technique with a group decision-making approach and BOCR for determining the criteria weights. They concluded natural gas to be the best alternative. The method was the basis of an MCDM model proposed by Kone and Burke (Kone & Buke, 2007) to identify the most suitable mixture of energy resources in Turkey for the generation of electric power. The BOCR method was applied to calculate the priority weights of the criteria, and it was observed that the highest priority was given to hydropower energy, while the lowest priority was ascribed to energy generated from oil. This MCDM technique was also used in the study conducted by Onut (Onut et al., 2008) for indicating the fuel resources that were most applicable for industries dealing with manufacturing, and he observed that the most widely used alternative was electric power, while the least was fuel-oil energy. ANP was used to compute the criteria weights for a methodology consisting of an amalgamation of BOCR and a model balancing scorecard, as presented by Shiue and Lin (Shiue & Lin, 2012), indicating the most applicable classification of recycling strategy in the industries dealing with solar power. At the end of the study, the best strategy determined was 'in-house.' Kabak and Dagdeviren (Kabak & Dagdeviren, 2014) combined ANP and BOCR methods to present an MCDM model which was used to classify numerous criteria in order to assess five alternatives of energy resources. The study concluded that energy generated from hydropower is the finest energy alternative.

11.4 TECHNIQUE FOR ORDER PREFERENCE BY SIMILARITY TO AN IDEAL SOLUTION (TOPSIS)

The Technique for Order Preference by Similarity to an Ideal Solution, one of the original dynamic techniques for solving MCDM issues, was created by Hwang and Yoon (Hwang & Yoon, 1981). The strategy depends on the rule that the selected option ought to have the farthest Euclidean distance from the Negative Ideal Solution (NIS), and the most limited from the Positive Ideal Solution (PIS). The more explicit solution that augments the advantage criteria and limits the cost criteria will be chosen as the best (Zouggari & Benyoucef, 2012). TOPSIS can be is versatile and can be used to supplant AHP in the process of positioning other options. At the end of the day, AHP is regularly utilized to relegate the heaviness of the determination criteria, while TOPSIS is applied to organize the other choice options.

Boran, [Boran et al., 2013] analyzed multiple criteria applying the TOPSIS technique for a comparative analysis of the contemporary sources of energy with nuclear power for the generation of electricity in the region of Turkey. These alternatives were analyzed with respect to various criteria, such as carbon dioxide emission, efficiency, acceptability, and total costs on the basis of various scenarios, and it was concluded that hydropower is the best alternative, followed by nuclear power, as evaluated for three scenarios. TOPSIS was utilized by Brand and Missaoui (Brand & Missaoui, 2014) to evaluate five different energy scenarios under economic, security of supply, socio-economic, and ecological criteria in determining that the most acceptable scenario selected for Tunisia was the diversification of renewable energy sources. (Afsordegan et al., 2014) compared qualitative TOPSIS and a non-compensatory outranking MCDM method to present a multi-criteria analysis framework for the selection of wind farm locations in Catalonia. At the end of the study, it was observed that both of these methods have similar performance qualifications. An extended version of the TOPSIS method, along with the weighted sum processes, was utilized (Alidrisi & Al-Sasi, 2017) to present an analysis of multiple energy-planning criteria for risk-prioritizing in the development of projects involving tidal energy. It was studied that prioritizing risks would create conflict from the academy and industry with a TOPSIS-based MCDM approach which intended to rank the countries that fall under G20 with respect to power generation from various sources of renewable energy. It was concluded that Germany and France were ranked the first for two distinct situations assessing the respective safe and unsafe energy resource as the nuclear power.

The technique of the TOPSIS strategy contains the following gradations.

Gradation One: A decision matrix is established by identifying the respective case's criteria and alternatives:

$$D = \begin{array}{c} \\ A_1 \\ \vdots \\ A_i \\ \vdots \\ A_m \end{array} \begin{array}{c} \begin{array}{ccccc} C_1 & \cdots & C_j & \cdots & C_n \end{array} \\ \begin{bmatrix} x_{11} & \cdots & x_{12} & \cdots & x_{1n} \\ \vdots & & \vdots & & \vdots \\ x_{i1} & \cdots & x_{ij} & \cdots & x_{in} \\ \vdots & & \vdots & & \vdots \\ x_{m1} & \cdots & x_{mj} & \cdots & x_{mn} \end{bmatrix} \end{array} \tag{11.6}$$

$$W = [w_1 \cdots w_j \cdots w_n]$$

In this case, multiple alternatives are depicted along with their respective efficacy rating, all of which are depicted in the form of a decision matrix.

Gradation Two: This grade involves normalisation of the decision matrix by using the given Equation 11.7:

$$R_{ij} = \overline{x_{ij}} / \sqrt{\sum_{k=1}^{n} x_{ik}^2}, \ (i = 1, 2, \dots m), \ (k = 1, 2, .. j .. n) \tag{11.7}$$

Gradation Three: This grade includes the construction of a fuzzy weighted normalized decision matrix.

Gradation Four: The final steps include determining the respective indexes for each alternative and their respective distances to finally calculate proximity.

11.5 VLSE KRITERIJUMSKA OPTIMIZCIJA I KAOMPROMISNO RESENJE (VIKOR)

This is a compromise ranking methodology that was developed in Serbia. It applies the approach of outranking on a limited set of selected alternatives with contradicting priorities and inequitable criteria. The legitimate accolades and credit of this MCDM approach belong to Duckstein and Opriovic, as quite lucidly stated by Opricovic and Tzeng (Opricovic & Tzeng, 2003). Primarily, it centers on asking and choosing the best from a number of alternatives and bargains the answers for an issue with clashing measures, which can help chiefs in arriving at an official choice. The trade-off arrangement is a plausible arrangement that is the closest to the ideal.

There are various stellar similarities between VIKOR and TOPSIS. However, the primary difference lies in the compromise that VIKOR is willing to make among the various selected alternatives and multiple priorities, while, on the contrary, TOPSIS aims to make the best solution by obtaining the shortest proximity among various available options (Opricovic & Tzeng, 2007).

11.6 ELIMINATION ET CHOIX TRADUISANT LA REALITÉ (ELECTRE)

A utility theory was employed by (Roy & Bouyssou, 1986) to examine the MCDM approaches and ranking distinctions on practical grounds. The outranking procedure of ELECTRE III was applied for the selection of plant locations producing nuclear energy, and it was analyzed how situations related to real-life could be the target for adopting outranking MCDM techniques. (Georgopoulou et al., 1997) applied the tri-version of the same technique for the selection of guidelines under the purview of renewable energy in the Greek islands in order to elucidate the requirement of evaluation in various criteria, except for the minimization of expenditure for problems related to policy-making related to local energy. For the planning of energy resources, the study estimated eight strategies, out of which two were determined as the most suitable. The technique was also applied along with the utilisation of a case study in Sardinia island in ranking three different scenarios and evaluating a plan of action to deal with the combination of renewable energy resources at the local level, as proposed in an MCDM approach by (Beccali et al., 2003). The most appropriate alternative scenario chosen was the reduction in the use of fossil fuels.

(Karagiannidis et al., 2018) intended to indicate the renewable energy alternatives that were possible to be inserted into an insular system for the production of electric power by applying the ELECTRE III method. The options were assessed on the basis of technical, social, environmental, and financial purview, and the results suggested inefficiency and high costs. Wind power was categorized as the best alternative for providing electricity. The ELECTRE TRI technique was conducted by (Karakosta et al., 2009) to rank the sustainable energy sources for the generation of electricity in a few selected countries. (Catalina et al., 2011) used the ELECTRE III procedure for multisource of systems in the selection and design of appropriate alternatives and obtained reasonable results by applying a case study to this developed approach. Three scenarios were analyzed for the prediction of the development of the sector related to energy resources in the region of Belgrade, by (Grujic et al., 2014), namely optimistic, realistic, and BAU. The outranking procedure of ELECTRE was used as an MCDM approach to obtain the best option for each of the situations by assessing the sources of energy in various criteria.

11.7 FUZZY LOGIC

The concept or the idea of 'fuzzy logic' was primarily developed and coined in 1965 (Zadeh, 1965) when creators eventually realized that decision-making is not a notional concept, but rather a pragmatic practice that is to be utilized in real-world problems where tasks are far away from the ideal. Furthermore, in reality, there are no absolute answers or 100% guarantees. Instead, more probabilities and uncertainties need to be approximated, calibrated, and quantized while making a logically and mathematically accurate decision. This is when the concept of fuzzy or uncertain logic gestated, by recognizing the uncertainties that one encounters in real-world decision-making and analyzing these for calculated risks. This concept of fuzzy logic developed and was talked about for more than 20 years in the scholar society around the globe. Mathematically, a fuzzy set is a head set or a super set of values of the type Boolean bearing relatively inaccurate information. Fuzzy logic primarily converts the levels of uncertainty which are considered as inputs in the form of literals, such as 'definite,' 'likely,' 'average,' 'unlikely,' and 'impossible,' to quantified mathematical values which can be calibrated, analyzed, and used in operations for further development.

To further include the concept of uncertainty and imprecise reasoning and insight gathered from approximation, fuzzy rules make for comparatively more decorous frameworks. It considers the concept that each value and insight needs to be deliberated and utilized in the edification and development of the framework. While only a few of the deliberated insights are perceived as complete or, in other words, most of the discerned perceptions and facts considered are not 'pure' or possessing probity, these are thus, considered partially. Very few attributes are considered 'pure' enough to be completely examined. This way of partial and complete consideration of various available attributes is known as the 'degree of membership.' The real-world usage and optimisation of complete fuzzy logic are generally confined to the domain of extreme business analytics, while partial fuzzy logic is largely applicable in variegated fields, like the development and up-gradation of Artificial Intelligence and expert systems (Pai et al., 2003). Moreover, a number of other problems have been used in multiple areas in disparate domains, like solving transportation issues by providing solutions to problems of transshipment hub port selection by deliberating the diverse attributes via fuzzy logic. Financial assessments are also one of the areas of fuzzy concept applications as applied to address the pecuniary evaluation of various companies in the shipping business and in terms of examining their lucrative nature by deliberating their literals and converting these into weights (Wang et al., 2014). This has been used as amalgamated with AHP (Chao & Lin, 2011), TOPSIS (Yeh & Chang, 2009), VIKOR (Kaya et al., 2010), ELECTRE (Sevkli, 2010), and PROMETHEE (Shirinfar & Haleh, 2011) in the previous decade.

11.8 CONCLUSION

A comprehensive analysis is presented in the given research publication, where each evaluated approach is reviewed on the basis of their individual primordial framework efficiency, application, and future scope. This chapter elucidates the basics of MCDM methodologies to provide a comprehensive and detailed understanding to the neophyte reader. The aim of this chapter is to not only introduce the basic concepts and pillar methodologies to the reader, but to also grant the user complete knowledge to decide which technique can be used via an in-depth understanding. The presented study successfully evaluates the various state-of-the-art classic Multi-Criteria Decision-Making techniques by elucidating their origin, the mathematics behind them, the various concepts, paradigms, merits, demerits, and disparate applications. This enables the readers to understand, comprehend, and analyze various MCDM approaches to choose the most appropriate technique that suits the current set and scenario of problems.

REFERENCES

Afsordegan, A., Sanchez, M., Agell, N., Aguado, J.C., & Gamboa, G. (2014). A comparison of two MCDM methodologies in the selection of a windfarm location in Catalonia. In L. Museros, O. Pujol, & N. Agell, (Eds.) *Frontiers in Artificial Intelligence Research and Development* (pp. 227–236). IOS Press.

Akash, B.A., Mamlook, R., & Mohsen, M.S. (1999). Multi-criteria selection of electric power plants using analytical hierarchy process. *Electric Power Systems Research*, *52*(1), 29–35.

Akbari, N., Irawan, C.A., Jones, D.F., & Menachof, D. (2017). A multi-criteria port suitability assessment for developments in the offshore wind industry. *Renew Energy*, *102*, 118–133.

Akyuz, E., Karahalios, H., & Celik, M. (2015). Assessment of the maritime labour convention compliance using balanced scorecard and analytic hierarchy process approach. *Maritime Policy & Management*, *42*(2), 145–162.

Al Garni, H., Kassem, A., Awasthi, A., Komljenovic, D., & Al-Haddad, K. (2016). A multi-criteria decision making approach for evaluating renewable power generation sources in Saudi Arabia. *Sustainable Energy Technologies and Assessments*, *16*, 137–150.

Alidrisi, H., & Al-Sasi, B.O. (2017). Utilization of energy sources by G20 countries: A TOPSIS-BASED approach. *Energy Sources, Part B: Economics, Planning, and Policy*, *12*(11), 964–970.

Aras, H., Erdogmus, S., & Koc, E. (2004). Multi-criteria selection for a wind observation station location using analytic hierarchy process. *Renewable Energy*, *29*(8), 1383–1392.

Barba-Romero, S., & Pomerol, J.C. (2000). *Multicriterion decision in management: Principles and practice*. Operations Research Management Science, Massachusetts.

Beccali, M., Cellura, M., & Mistretta, M. (2003). Decision-making in energy planning. Application of the electre method at regional level for the diffusion of renewable energy technology. *Renew Energy*, *28*(13), 2063–2087.

Beşikçi, E.B., Kececi, T., Arslan, O., & Turan, O. (2016). An application of fuzzy-AHP to ship operational energy efficiency measures. *Ocean Engineering*, *121*, 392–402.

Blanco, G., Amarilla, R., Martinez, A., Llamosas, C., & Oxilia, V. (2017). Energy transitions and emerging economies: A multi-criteria analysis of policy options for hydropower surplus utilization in Paraguay. *Energy Policy*, *108*, 312–321.

Bojesen, M., Boerboom, L., & Skov-Petersen, H. (2015). Towards a sustainable capacity expansion of the Danish biogas sector. *Land Use Policy*, *42*, 264–277.

Boran, F.E., Etoz, M., & Dizdar, E. (2013). Is nuclear power an optimal option for electricity generation in Turkey. *Energy Source*, *8*(4), 382–390.

Brand, B., & Missaoui, R. (2014). Multi-criteria analysis of electricity generation mix scenarios in Tunisia. *Renewable and Sustainable Energy Reviews*, *39*, 251–261.

Carlos Perez-Mesa, J., Galdeano-Gomez, E., & Salinas Andujar, J.A. (2012). Logistics network and externalities for short sea transport: An analysis of horticultural exports from southeast Spain. *Transport Policy*, *24*, 188–198.

Catalina, T., Virgone, J., & Blanco, E. (2011). Multi-source energy systems analysis using a multi-criteria decision aid methodology. *Renew Energy*, *36*(8), 2245–2252.

Chao, S.L., & Lin, Y. (2011). Evaluating advanced quay cranes in container terminals. *Transp Res Part E-Logist Transp Rev*, *47*(4), 432–445.

Chen, T.L., & Lin, K.L. (2006). Complementing AHP with habitual domains theory to identify key performance indicators for service industry. In: *IEEE international conference on service operations and logistics and informatics*, pp. 84–89.

Chou, C.C., & Ding, J.F. (2016). An AHP model for the choice of ship flag: A case study of Tawanese shipowners. *International Journal of Maritime Engineering*, *185*, A61–A68.

Erdogmus, S., Aras, H., & Koc, E. (2006). Evaluation of alternative fuels for residential heating in Turkey using analytic network process (ANP) with group decision-making. *Renewable and Sustainable Energy Reviews*, *10*(3), 269–279

Georgopoulou, E., Lalas, D., & Papagiannakis, L. (1997). A multicriteria decision aid approach for energy planning problems: The case of renewable energy option. *European Journal of Operational Research*, *103*(1), 38–54.

Grujic, M., Ivezic, D., & Zivkovic, M. (2014). Application of multi-criteria decision making model for choice of the optimal solution for meeting heat demand in the centralized supply system in Belgrade. *Energy*, *67*, 341–350.

Haddad, B., Liazid, A., & Ferreira, P. (2017). A multi-criteria approach to rank renewables for the Algerian electricity system. *Renewable Energy*, *107*, 462–472.

Hwang, C.L., & Yoon, K. (1981). *Multiple attributes decision making methods and applications*. Springer, Berlin.

Kabak, M., & Dagdeviren, M. (2014). Prioritization of renewable energy sources for Turkey by using a hybrid MCDM methodology. *Energy Conversion and Management*, *79*, 25–33.

Kablan, M.M. (2004). Decision support for energy conservation promotion: An analytic hierarchy process approach. *Energy Policy*, *32*(10), 1151–1158.

Kagazyo, T., Kaneko, K., Akai, M., & Hijikata, K. (1997). Methodology and evaluation of priorities for energy and environmental research projects. *Energy*, *22*(2–3), 121–129.

Karakosta, C., Doukas, H., & Psarras, J. (2009). Directing clean development mechanism towards developing countries' sustainable development priorities. *Energy for Sustainable Development*, *13*(2), 77–84.

Karagiannidis, A., et al. (2018). Application of the multi-criteria analysis method Electre III for the optimisation of decentralised energy systems. *Omega*, *36*(5), 766–776.

Kaya, T., & Kahraman, C. (2010). Multicriteria renewable energy planning using an integrated fuzzy VIKOR and AHP methodology: The case of Istanbul. *Energy*, *35*(6), 2517–2527

Kone, A.C., & Buke, C. (2007). An analytical network process (ANP) evaluation of alternative fuels for electricity generation in Turkey. *Energy Policy*, *35*(10), 5220–5228.

Kumar, R., Kumar, V., & Singh, S. (2017a). Modeling and analyzing the impact of lean principles on organizational performance using ISM approach. *Journal of Project Management*, *2*, 37–50.

Kumar, R., Kumar, V., & Singh, S. (2017b). Work culture enablers: Hierarchical design for effectiveness & efficiency. *International Journal of Lean Enterprise Research (IJLER)*, *2*(3), 189–201.

Kumar, K., Dhillon, V.S., Singh, P.L., & Sindhwani, R. (2019). Modeling and analysis for barriers in healthcare services by ISM and MICMAC analysis. In M. Kumar, R. Pandey, & V. Kumar (Eds.) *Advances in Interdisciplinary Engineering*. Lecture Notes in Mechanical Engineering (pp. 501–510). Springer, Singapore.

Malkawi, S., Al-Nimr, M., & Azizi, D. (2017). A multi-criteria optimization analysis for Jordan's energy mix. *Energy*, *127*, 680–696.

Mittal, V.K., Sindhwani, R., Singh, P.L., Kalsariya, V., & Salroo, F. (2018). Evaluating significance of green manufacturing enablers using MOORA method for Indian manufacturing sector. In Proceedings of the *International Conference on Modern Research in Aerospace Engineering* (pp. 303–314). Springer, Singapore.

Mittal, V.K., Sindhwani, R., Shekhar, H., & Singh, P.L. (2019). Fuzzy AHP model for challenges to thermal power plant establishment in India. *International Journal of Operational Research*, *34*(4), 562–581.

Onut, S., Tuzkaya, U.R., & Saadet, N. (2008). Multiple criteria evaluation of current energy resources for Turkish manufacturing industry. *Energy Conversion and Management*, *49*(6), 1480–1492.

Opricovic, S., & Tzeng, G.H. (2003). Multicriteria expansion of a competence set using genetic algorithm. In Tanino, T., Tanaka, T., & Inuiguchi, M. (Eds.) *Multi-objective programming and goal-programming: Theory and applications* (pp. 221–226). Springer, Berlin/Heidelberg.

Opricovic, S., & Tzeng, G.H. (2007). Extended VIKOR method in comparison with outranking methods. *European Journal of Operational Research*, *178*(2), 514–529.

Pai, R.R., Kaflepalh, V.R., Caudill, R.J., Zhou, M.C., & Leee, I. (2003). Methods toward supply chain risk analysis. In *2003 I.E. international conference on systems, man and cybernetics, vols 1–5, conference proceedings*, pp. 4560–4565.

Pohekar, S.D., & Ramachandran, M. (2004). Application of multi-criteria decision making to sustainable energy planning—A review. *Renewable and Sustainable Energy Reviews*, *8*(4), 365–381.

Rosso, M., Bottero, M., Pomarico, S., La Ferlita, S., & Comino E. (2014). Integrating multicriteria evaluation and stakeholders analysis for assessing hydropower projects. *Energy Policy*, *67*, 870–881.

Roy, B., & Bouyssou, D. (1986). Comparison of two decision-aid models applied to a nuclear power plant siting example. *European Journal of Operational Research*, *25*(2), 200–215.

Saaty, T.L. (1980). *The analytic hierarchy process*. McGraw-Hill, New York.

Saaty, T.L. (1990). How to make a decision – The analytical hierarchy process. *European Journal of Operational Research*, *48*(1), 9–26.

Sagbansua, L., & Balo, F. (2017). Decision making model development in increasing wind farm energy efficiency. *Renewable Energy*, *109*, 354–362.

Sahin, B., & Senol, Y.E. (2015). A novel process model of marine accident analysis by using generic fuzzy-AHP algorithm. *Journal of Navigation*, *68*, 162–183.

Sevkli, M. (2010). An application of the fuzzy ELECTRE method for supplier selection. *International Journal of Production Research*, *48*(12), 3393–3405.

Shirgholami, Z., Zangeneh, S.N., & Bortolini, M. (2016). Decision system to support the practitioners in the wind farm design: A case study for Iran mainland. *Sustainable Energy Technologies and Assessments*, *16*, 1–10.

Shirinfar, M., & Haleh, H. (2011). Supplier selection and evaluation by fuzzy multi-criteria decision making methodology. *International Journal of Industrial Engineering and Production Research*, *22*(4), 271–280.

Shiue, Y.C., & Lin, C.Y. (2012). Applying analytic network process to evaluate the optimal recycling strategy in upstream of solar energy industry. *Energ Buildings*, *54*, 266–277.

Sii, H.S., & Wang, J. (2003). A design-decision support framework for evaluation of design options/proposals using a composite structure methodology based on the approximate reasoning approach and the evidential reasoning method. *Proceedings of the Institution of Mechanical Engineers, Part E: Journal of Process Mechanical Engineering*, *217*(E1), 59–76.

Sindhu, S., Nehra, V., & Luthra, S. (2017). Solar energy deployment for sustainable future of India: Hybrid SWOC-AHP analysis. *Renewable and Sustainable Energy Reviews*, *72*, 1138–1151.

Shanker, K., Shankar, R., & Sindhwani, R. (Eds.). (2019). *Advances in industrial and production engineering*: *Select proceedings of FLAME 2018*. Springer.

Sindhwani, R., & Malhotra, V. (2017). A framework to enhance agile manufacturing system: A Total Interpretive Structural Modelling (TISM) approach, *Benchmarking: An International Journal*, *24*(2), 467–487. https://doi.org/10.1108/BIJ-09-2015-0092

Sindhwani, R., Mittal, V.K., Singh, P.L., Kalsariya, V., & Salroo, F. (2018). Modelling and analysis of energy efficiency drivers by fuzzy ISM and fuzzy MICMAC approach. *International Journal of Productivity and Quality Management*, *25*(2), 225–244.

Sindhwani, R., Singh, P.L., Chopra, R., Sharma, K., Basu, A., Prajapati, D.K., & Malhotra, V. (2019). Agility evaluation in the rolling industry: A case study. In K. Shanker, R. Shanker, & R. Sindhwani (Eds.). *Advances in Industrial and Production Engineering*. Lecture Notes in Mechanical Engineering (pp. 753–770). Springer, Singapore.

Sindhwani, R., Singh, P.L., Iqbal, A., Prajapati, D.K., & Mittal, V.K. (2019). Modeling and analysis of factors influencing agility in healthcare organizations: An ISM approach. In K. Shanker, R. Shanker, & R. Sindhwani (Eds.). *Advances in Industrial and Production Engineering*. Lecture Notes in Mechanical Engineering (pp. 683–696). Springer, Singapore.

Talaei, A., Ahadi, M.S., & Maghsoudy, S. (2014). Climate friendly technology transfer in the energy sector: A case study of Iran. *Energy Policy*, *64*, 349–363.

Tsai, W., Leu, J., Liu, J., Lin, S., & Shaw, M. (2010). A MCDM approach for sourcing strategy mix decision in IT projects. *Expert Systems with Applications*, *37*(5), 3870–3886.

Ugboma, C., Ibe, C., & Ogwude, I.C. (2004). Service quality measurements in ports of a developing economy: Nigerian ports survey. *Managing Service Quality: An International Journal*, *14*(6), 487–495.

Ulutas, B.H. (2005). Determination of the appropriate energy policy for Turkey. *Energy*, *30*(7), 1146–1161.

Ung, S.T., Williams, V., Chen, H.S., Bonsall, S., & Wang J. (2006). Human error assessment and management in port operations using fuzzy AHP. *Marine Technology Society Journal*, *40*(1), 73–86.

Vaidya, O.S., & Kumar, S. (2006). Analytic hierarchy process: An overview of applications. *European Journal of Operational Research*, *169*(1), 1–29.

Wang, J.J., Jing, Y.Y., Zhang, C.F., Zhang, X.T., & Shi G.H. (2008). Integrated evaluation of distributed triple-generation systems using improved grey incidence approach. *Energy*, *33*(9), 1427–1437.

Wang, Y.J., & Lee, H.S. (2010). Evaluating financial performance of Taiwan container shipping companies by strength and weakness indices. *International Journal of Computer Mathematics*, *87*, 38–52.

Wang, Y., Jung, K.A., Yeo, G.T., & Chou, C.C. (2014). Selecting a cruise port of call location using the fuzzy-AHP method: A case study in East Asia. *Tourism Management*, *42*, 262–270.

Wang, P., Li, Y., Wang, Y.H., & Zhuo, Z.Q. (2015). A new method based on TOPSIS and response surface method for MCDM problems with interval numbers. *Mathematical Problems in Engineering*, *2015*, 11, http://dx.doi.org/10.1155/2015/938535

Xiaohua, W., & Zhenmin, F. (2002). Sustainable development of rural energy and its appraising system in China. *Renewable and Sustainable Energy Reviews*, *6*(4), 395–404.

Yeh, C.H., & Chang, Y.H. (2009). Modeling subjective evaluation for fuzzy group multicriteria decision making. *European Journal of Operational Research*, *194*(2), 464–473.

Yu, R., & Tzeng, G.H. (2006). A soft computing method for multi-criteria decision making with dependence and feedback. *Applied Mathematics and Computation*, *180*, 63–75.

Zadeh, L.A. (1965). Fuzzy sets. *Information and Control*, *8*(3), 338–353.

Zimmermann, H.J. (2000). An application-oriented view of modeling uncertainty. *European Journal of Operational Research*, *122*(2), 190–198.

Zouggari, A., & Benyoucef, L. (2012). Simulation based fuzzy TOPSIS approach for group multi-criteria supplier selection problem. *Engineering Applications of Artificial Intelligence*, *25*(3), 507–519.

Index

A

Automotive Brake, 23
Analytic Hierarchy Process (AHP), 88–9, 191
  Network Process (ANP), 89, 195

B

Brake Friction, 24
Business Strategy, 179

C

Combined Method Style Approach, 92–3
Conflict Management, 3, 8
Criteria Weight, 30, 47, 49
Criteria Importance Through Inter-Criteria Correlation (CRITIC), 25
Customer Feedback Management, 4, 8

D

Decision-Making, 83, 86, 139
  Trial and Evaluation Laboratory (DEMATEL), 94, 107–8, 111
Decision Matrix (DM), 28

E

Economic, 69
ELimination Et Choix Traduisant la REalité (ELECTRE), 90–1, 198
Enterprise-Wide Integration of Learning, 4, 7
Entropy, 23, 29, 43, 125
External Collaborations, 5, 6

F

Facility Management, 2, 14
Fuzzy Analytic Hierarchy Process (FAHP), 1, 39, 155
  DEMATEL Approach, 1, 07, 111
  Logic, 1, 44, 199
  Number, 144–5, 165–6
  Set, 121

G

GDP, 66, 78

H

Healthcare Sector, 2
Hirerchichal structure, 157

I

Innovation, 69
Inter-Connectivity Management, 4, 6
Interperative Structural Modelling (ISM), 67–9
IT Tools/Technologies Adoption, 5, 7

L

Leadership, 69
Linguistic Intuitionistic Fuzzy Sets, 1, 21, 125

M

Managerial Implication, 79
Mathematical Modeling, 139
Market Awareness, 4, 9
Methodology, 5
MICMAC, 10, 76
Module Efficiency, 1, 57, 168
Multi Attribute Decision Making (MADM), 44
  Criteria Decision Analysis, 1, 79, 181
  Criteria Decision Making (MCDM), 23, 47, 67, 83, 129, 155, 189

O

Organisational Structure Design, 3, 6

P

Patient Assumption and Hospital Discernment, 5, 8
Preliminaries, 123
Preference Ranking Organization Method for Enrichment of Evaluations (PROMETHEE), 91–2

R

Renewable Energy, 83, 87

S

Small & Medium Scale Enterprises (SME's), 67, 68
Solar Panel Module, 155
Strategy, 18, 01, 82, 183
  Decision-Making, 182
Sustainable Human Resource Management, 1, 07, 109
  Supply Chain Management, 109
System Thinking, 186

T

Talent Management, 3, 7
Technique for Order Preference by Similarity to an Ideal Solution (TOPSIS), 43, 48, 51, 58, 90, 196
Triangular Fuzzy Number, 1, 12, 144
Total Interperative Structural Modelling (TISM), 7

V

Vlse Kriterijumska Optimizcija I Kaompromisno Resenje (VIKOR), 23, 31, 43, 48, 50, 61, 198

W

Weight, 27, 28

Y

Yield, 14, 51, 56, 157